BATHYMETRY AND ITS APPLICATIONS

Edited by **Philippe Blondel**

Bathymetry and Its Applications
Edited by Philippe Blondel

Published by InTech
Janeza Trdine 9, 51000 Rijeka, Croatia

Publishing Process Manager Maja Bozicevic

Technical Editor Teodora Smiljanic

Cover Designer InTech Design Team

Additional hard copies can be obtained from orders@intechopen.com

Bathymetry and Its Applications, Edited by Philippe Blondel
p. cm.
ISBN 978-953-307-959-2

For EU product safety concerns:
IN TECH d.o.o., Prolaz Marije Krucifikse Kozulić 3, 51000 Rijeka, Croatia,
info@intechopen.com or visit our website at intechopen.com.

Meet the editor

Dr Philippe Blondel has written several books about sonar and its applications, backed by experience in all oceans (except the Antarctic). He is the Deputy Director of the Centre for Space, Atmosphere and Oceanic Science (CSAOS), and a Senior Lecturer in the Department of Physics at the University of Bath (UK). He previously worked at the Southampton Oceanography Centre (SOC), and the Institute of Oceanographic Sciences, Deacon Laboratory (IOSDL) in the UK, and the School of Oceanography, University of Washington in the USA. His research group is active in developing new acoustic sensors, designing new processing techniques, and applying them to habitat mapping, marine geology (e.g. cold-water coral reefs and tsunami-generating areas) and the effects of climate change on marine environments.

Contents

Preface

More than two thirds of the Earth is covered with water, and most of this is inaccessible to direct, visual observations. It is a cliché, but true nonetheless, to admit that we know more about Venus, a planet of the size of the Earth but millions of kilometers away, than about our own shores and seas, just a few kilometers away. Recent decades have seen a revolution in this knowledge, as we have slowly but steadily measured more and more of the water-covered surface of our own planet. Measuring the topography of this surface has been crucial in revolutionizing our understanding of the evolution of our planet, from crustal processes to sea level changes, and bathymetry is now a tool of choice when mapping underwater areas and looking at their monitoring and management. Entitled *Bathymetry and its Applications*, this book has been divided into two sections: Measuring Bathymetry and Applications of Bathymetry. This separation is somewhat arguable, as the measurements inevitably lead to applications, and successful applications forcibly rely on a good understanding of how the measurements were taken.

The first section, Measuring Bathymetry, starts with an article by Dr Bishwajit Chakraborty and Dr William Fernandes on acoustic echosounders, the traditional workhorse of bathymetry measurements. Showing the differences but also the complementarity between single-beam and multi-beam echosounders, the article explains the different steps of processing, highlighting them with examples taken from the authors' experience around the Indian Ocean, and proposing interesting new approaches for data analyses. Backed with a short but complete bibliography, this article illustrates clearly the traditional approach to bathymetry measurements. The following article, by Dr Julian Vrbancich, presents another tool for measuring bathymetry, with airborne electromagnetic sensors. Necessarily limited by the underlying physics to shallower depths, these emerging systems provide excellent-quality data on the seabed and its sub-surface. This article presents a wealth of examples, taken from all around Australia by the author and colleagues. It plainly shows the advantages of airborne electromagnetic sensors in clearly specified areas and depths, and the extensive bibliography provides many pointers for future developers of new sensors as well as future and current users of existing systems. The final article in this section, by Dr Maged Marghany, shows the emerging use of radar sensors to derive bathymetry from either airborne or spaceborne radars. Comparisons with other studies show the potential of these techniques in assisting traditional acoustic sensors by providing a rapid view of coastal bathymetry in shallow regions.

The second section, Applications of Bathymetry, starts with an article by Professor Martin Hovland on the geomorphology and nature of seabed seepage processes. Professor Hovland has been a key player in the study of gas seepage over the last decades, and several structures on the deep seabed have been named after him. In his article, he shows the primordial use of bathymetry in studying gas seeps, pockmarks and cold-water coral reefs. This article and the underlying research shows how bathymetry measurements, carefully combined with other physical and chemical measurements, can be of great use to economic geologists, biologists, and chemists as well as marine geologists in general. The second article in this section, submitted by Dr José de Anda and including several colleagues from Mexico, the US, and Germany, shows the extension of bathymetry to the use of Digital Elevation Models for large bodies of water on land. This article focuses on large tropical reservoirs, of great importance for river management and water resources monitoring, and sustainable use. The third article, by Dr Cynthia Fernandes Pinto da Luz, introduces the use of palynology in bathymetry measurements, to inform the acquisition of bathymetry data (when constrained by time and/or costs), and to add to the interpretation of the resulting datasets.

The research presented throughout the book is of very high quality, combining several disciplines and coming from very diverse regions around the world (namely India, Australia, Malaysia, Norway, Mexico, USA, Germany, and Brazil). Drawing on the efforts of other researchers from many other countries, and true to the InTech motto of "Open science, open minds", it aims at presenting readers everywhere with some of the latest accomplishments in research on bathymetry and its applications. Backed with good references to other work by the respective authors, and similar studies by other scientists, this book will surely provide an inspiration to all developers of bathymetry sensors and users of bathymetry applications. Similar open-access InTech books have been read thousands of times, and individual articles downloaded hundreds of times, in the few years since their publication. This bodes very well for *Bathymetry and its Applications*.

It is six months to the day that I was invited by Mr Aleksandar Lazinica, CEO of InTech Open Access Publisher, to edit this book. The different authors had already been contacted, and they had sent short abstracts of the research they wanted to present. After selection, the best ones were selected for inclusion in this book. The rapid publication schedule is evidence of the eagerness of the authors to present their research to a wider community, actively and positively engaging with the revisions always necessary in presenting peer-reviewed articles. It is also a proof of the hard work of InTech staff. The presentation of this book would not have been possible without the persistent efforts of everyone at InTech, in particular Ms Maja Bozicevic,

Publishing Process Manager, but also the Technical Editor and the Design team. All contributed, in varied but similarly efficient ways, towards making this book a reality. All deserve the gratitude of the authors, but also of the many readers who will be able to access these research results for free.

Philippe Blondel
CGeol FGS FIOA PhD
Deputy Director, Centre for Space, Atmosphere and Oceanic Science
University of Bath, Bath,
UK

Part 1

Measuring Bathymetry

1

Bathymetric Techniques and Indian Ocean Applications

Bishwajit Chakraborty and William Fernandes
National Institute of Oceanography (Council of Scientific & Industrial Research)
India

1. Introduction

Around 100 years ago, the ocean bottom was thought to be flat and featureless. But with the advent of modern echo sounding techniques the picture has significantly changed. The aim of shallow water bathymetry (measurement and charting of the sea bottom) is to provide navigational safety whereas deep ocean survey is generally of an exploratory nature. Rough mountainous terrain, including the mid-ocean ridge system, is known to cover larger portions of the seabed. In order to understand dynamic processes related to the shape of earth the seafloor bathymetric explorations are important along with the routine offshore explorations of mineral deposits, especially in the Exclusive Economic Zones (EEZ). The Earth works as an integrated system of interacting bio-geophysico-chemical processes which are influenced by the land topography and ocean bathymetry. Erosion and sedimentation rates are much lower in the ocean than on land. However, detailed bathymetry reveals the morphology and geological history. The understanding of the related geological and geophysical parameters which shape the ocean floor, are also essential for living and non-living resource estimation. The ocean floor acts as an interface between the oceanic lithosphere and hydrosphere, and the interaction (i.e., exchange of mass and energy) zone for the processes are provided by the boundary layers. Evidences of the interaction effects in terms of the geomorphology can be seen in the seafloor sediment of ripple marks and bioturbation. Similarly, geological evidences can be seen in the development of manganese nodules and forms of past and present submarine volcanism. Generally, the two basic processes which shape the seafloor are known as endogenic and exogenic e.g. Seibold & Berger (1993). The large seafloor features (seamount, ridge crest, valley etc) related to the plate tectonics are due to the endogenic processes i.e., those deriving their energy from the earth interiors. Small scale features (ripple /abyssal plain etc.) due to erosion as well as deposition of the sediments are attributed to the exogenic processes i.e. those driven by the Sun - a major parameter controls physical processes such as climate temperature and wind waves etc. In turn, these processes control fine-scale deep seafloor morphology. Two third of the earth surface i.e. 362 million square km (70 %) is covered by the ocean. In order to understand the seafloor various methods such as application of remote acoustic techniques (Lurton, 2002), seafloor photographic and geological sampling techniques are well established. Echo-sounding through use of hull mounted transducer became familiar during World War II. Advantage of this technique lies with the rapid depth data acquisition. Due to the improvement of the material science, the designing of the low cost but high quality transducers became more widespread during the year 1950. The high resolution single beam echo-sounder (higher frequency and narrower beam-width) was available in 1970. Besides single beam echo-sounding technique, which provides single data points beneath the sea, other high resolution remote acoustic techniques like side scan sonar and sub-bottom technique became popular. GLORIA side scan sonar (operating frequency: 6 kHz) as well as SEAMARC (operating frequency: 12 kHz) were extensively employed by USGS during the 1980's. Side scan sonar system offer seabed aerial view, however, they do not provide accurate depth information. This is primarily due to the phase measuring techniques applied to ensure higher coverage, and estimation of the depth near nadir regions is still contested. Nonetheless, the bathymetry of the side scan sonar systems based on the interferometric techniques such as TOBI (NERC, UK), DSL-120 (WHOI, USA) and Swath Plus (SEA Ltd, UK) were also used extensively. The system details are well covered in Blondel (2009).

Single beam echo sounder replaced the lead-line methods from the 1920s onwards and with this, continuous records of depths along track of the ship were available. Single beam echo sounding technique came into common usage around the 1950s which was only possible after the improvement of transducer technology. The spatial resolution of such systems is a function of the half power beam width, and the transducer beam widths were available around 30^0-60^0. In order to provide higher resolution, narrow beam (half power beam widths, 2^0-5^0) techniques were introduced much later. These systems became fully operational for real-time attitudes like roll pitch correction since the 1960s i.e., introduction of electronic stabilization. Beam steering techniques were used for such application. Using narrow single beam echo-sounding systems, seabed relief from meter to kilometer can be recorded continuously along the ship's track. But in order to generate a bathymetric map, several parallel profiles at short intervals are required. Multi-beam bathymetric system (Fig. 1), give bottom profiles which are correlated with respect to the single central track, and allows more reliable correlation of intersecting tracks in a single strip high density depth survey. In 1970, the multi-beam bathymetric system became commercially available with additional facilities like real-time computation and data storage capabilities. Apart from depth determination, this sounding system is useful for various scientific and survey objectives, such as geological survey to characterize the seabed, geotechnical properties and resource evaluation such as polymetallic nodule, ridge research, gas hydrate studies, habitat research etc. The high resolution and high density bathymetry data is found to be economical since it maintains higher depth accuracies and coverage. Multi-beam echo-sounder is a recent successor to single beam technique. Seabeam-multi-beam system was available in 1970's. Multi-beam technique utilizes multiple narrow beam transmission/ reception for a single transmission providing better seafloor coverage. Initial 16 beam Seabeam system had a limited coverage which was subsequently increased to 5-7 times of the centre beam depth. Numbers of beams were generally increased to have high-resolution beam-width varying within the 1.5°-2.2°. In India first multi-beam- Hydrosweep system was installed onboard Ocean Research vessel (ORV) Sagar Kanya (owned by the Ministry of the Earth Science, New Delhi), and it was operated by the National Institute of Oceanography during the year 1990 (Kodagali & Sudhakar, 1993**).** The Hydrosweep system used to form 59 receiving beams having 2.2° beam-widths which were covering the seafloor over twice the centre-beam depth. Use of modern computers provide faster signal processing techniques connected with position fixing equipments such as Differential Global Positioning System

(DGPS) which helps in collocating depth data with the beam position. Extensive bathymetric data acquisition and seafloor investigation was carried out under various scientific projects like: survey for polymetallic nodule, "ridge" i.e., mid-ocean ridge research etc. Recently, mapping program of the 2.2 million km^2 area of Exclusive Economic Zone (EEZ) of India using multi-beam system has been undertaken by the Ministry of the Earth Sciences, New Delhi, to which the Indian Institutes e.g., the National Institute of Oceanography, the National Institute of Ocean Technology, and the National Centre of Antarctic and Ocean Research are major contributors. Apart from acquiring the bathymetry data, multi-beam systems were modified to acquire backscatter data also. Backscatter information of the seafloor provide textural aspect of the seafloor i.e., idea about the seafloor sediment material and small scale roughness. This information along with the bathymetry for structural aspect is extremely useful to understand the seafloor. Backscatter strength values can match with the conventional side scan sonar system and offer side scan sonar data along with the bathymetry. Moreover, angular backscatter strength data can be utilized to estimate quantitative seafloor roughness parameters such as sea-water-floor interface as well as sediment volume roughness. Again, at NIO, modification in the Hydrosweep - multi-beam system was made in 1995. Significant seafloor studies are being carried out at NIO during the last two decades where shallow and deep water bathymetric surveys have become compulsory for seafloor studies. Various seafloor segmentation techniques to classify the seafloor were initiated. Numerical based inversion techniques along with the soft computational techniques were used (Chakraborty et al, 2003). Present study will endeavor to elucidate modern techniques to understand seafloor processes and illustrate them with appropriate examples around the Indian Ocean.

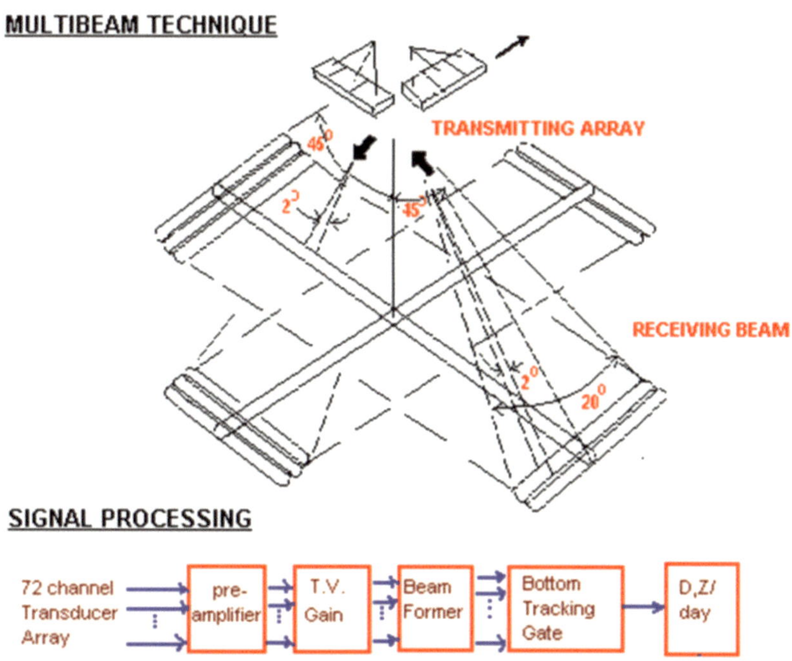

Fig. 1. Transducer arrays and signal processing systems used in multi-beam techniques. T.V. Gain, D and Z are the Time Varied Gain (TVG), depth and time display units respectively.

2. Indian Ocean bathymetry

The Exclusive Economic Zone (EEZ) of India is covered with a variety of minerals of economic interest. The EEZ extends up to 200 nautical miles from the coastline. The coastal state has the right to explore and exploit the resource and protect the environment. The non-living resources present in the EEZ around India constitute scope for the exploration of hydrocarbon, calcium carbonate and phosphorite etc. Apart from that, it has also suitable for hydrothermal mineral exploration in the Andaman back arc basin. Multi-beam bathymetric mapping is essential to understand the seabed morphological aspects of these areas in order to acquire scientific and engineering information from the point of view of exploitation. Multi-beam surveys conducted around western continental margin reveals the presence of many small to medium scale seabed topographic features which were not seen in the previous exploration using single beam echo-sounding systems. Some of the seafloor features will be covered in this article in terms of deep as well as shallow water areas.

2.1 Bathymetry of shallower areas

Small scale topographic prominences were seen during the echo-sounding surveys, those are formed of algal and/or oolitic limestone on the outer continental shelf off many tropical coasts. Similar features were observed during the course of multi-beam echo-sounding survey on the "Fifty Fathom Flat" off Mumbai, the west coast of India (Nair, 1975). The shelf width off Mumbai is 300 km. The outer 250 km of the shelf, have depth within the 65 to 100m. Small scale features such as pinnacles having heights generally varying from 1-2m, occasionally reaching a maximum height of 8m at some places along with the associations of trough like features. Progressive increase in the relief is observed as one proceeds from shallow to deep waters. In addition to these features, mound shaped protuberances are also varying from 2000-4000m. The height of such mound shape features are varying within the 6-8m. These features are prominently observed beyond the outer shelf of 80-85m, and shelf break in this area occurs at 95m water depth. Single beam bathymetry available on this margin showed reef-like features and large mounds, both made up of aragonite sands (Rao et al, 2003). The dimensions of these features needing detailed multi-beam survey. It is well known that during the Last Glacial Maximum (LGM) (18,000 ^{14}C yrs BP), the Glacio-eustatic sea level was at about 120-130 m below the present position (Fairbanks, 1989). Thereafter the sea level started rising largely due to the melting of glacial ice. The carbonate platform measuring about 28000 km^2, extending 4 degrees of latitude lies between the 60 to 110m water depths on the outer continental shelf of the northwestern margin of India. Although the platform lies off major rivers such as Narmada and Tapi, it contains <10% terrigenous material. The relic deposits on the carbonate platform are largely carbonate sands. It is interesting to know how carbonates developed when one considers the geographic setting (off major rivers), age of the sediments of the platform, and environmental conditions (intense monsoon) during the early Holocene (Rao & Wagle, 1997). The absence of terrigenous flux on the platform and continued carbonate growth until 7.6 kaBP, implies that the riverine flux either filled the inner shelf or diverted towards the south under the influence of a southwest monsoon current. Therefore, detailed high resolution multi-beam bathymetry survey should be carried out to understand whether this is biohermal or manifestation of physical processes during lower sea levels. This, in-turn helps to identify the factors inhibiting the terrigenous flux on to the platform and processes operative during lowered sea levels. Below, multi-beam bathymetric map of "Fifty Fathom Height" off

Tarapur, Mumbai is presented (Fig. 2). The water depth varies from 70-120m having vertical exaggeration of thirty providing the clear indication of small scale reef like features. The data was acquired using EM 1002 multi-beam system installed on-board Coastal Research Vessel (CRV) Sagar Sukti.

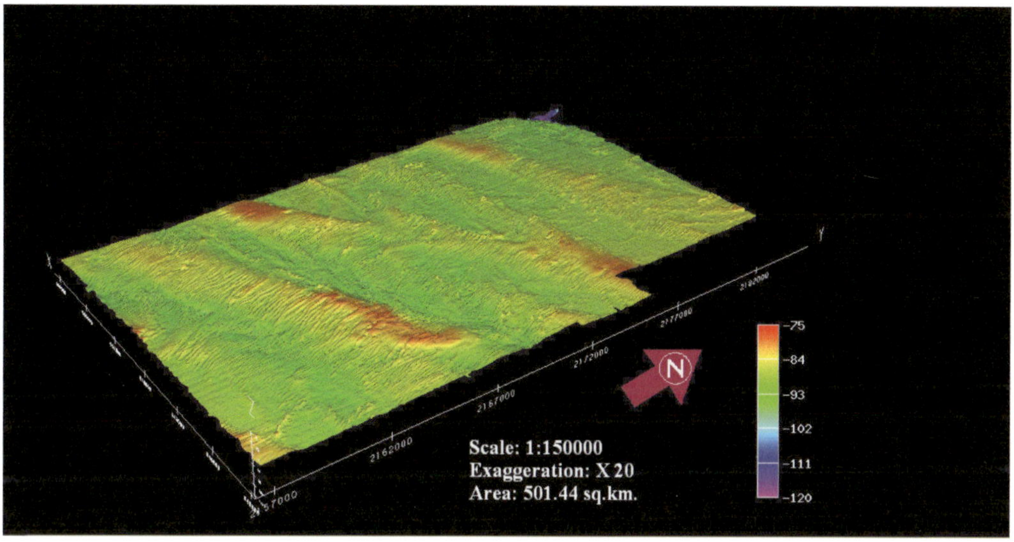

Fig. 2. Multi-beam bathymetric map of the "Fifty Fathom Flat" off Mumbai covering distances along northern and southern axes are 28.25 km and 17.75 km respectively.

On the western continental shelf of India, the middle and outer shelf environment, beyond 60 m depth, is uneven and is characterized by the presence of relict carbonate-rich sediments and a variety of limestones. Numerous prominent reefal structures between water depths of 38 and 136 m have been identified, having positive relief, which is believed to be due to the biohermal growth. These reefs occur seaward within the inner and outer shelf transition zone, as significant number of peaks, pinnacles and protuberances of different heights. The height varies from <1 m to 14 m. In shallow environments, often the reefs are buried below 5 - 10 m thick sediments. Bathymetry data of the western continental map is shown, which has also been acquired using EM 1002 system. Depth variation of 60-85m is shown off northern part of the Goa. Small-scale structures as discussed are prominently seen in the figure (Fig. 3) covering 197 km^2 area are presented with a vertical exaggeration of 10. Progressive narrowing of the shelf is observed in the western continental shelf. It is around 60m wide off Goa, and shelf break occurring at 130m water depth. Terrace like features are observed here, however, such features are not so prominent than the off Mumbai area.

Quasim & Nair (1978) discovered a living coral bank at 80m water depth, about 100km from the coast off Malpe (Karnataka state), west coast of India and named it as Gaveshani bank. The bank has a height of 42m, length of 2km and width of 1.66 km having area of 3.00 km^2 (Figure 4). Walls of this bank rise steeply from the seafloor. Sediments collected by grabs from the seafloor around the bank were silty sand, predominantly carbonate, consisting of shells of foraminifera, fragments of mollusks and corals. The radio carbon age of the sediments and rocks from the outer continental shelf is between the 9000 and 11000 yr. This period corresponds to Holocene when sea level began to rise or when it was in transgressive state. Living corals were acquired from the bank and five major species were identified. Coral growth in the area might have started in the Pleistocene when sea level was low.

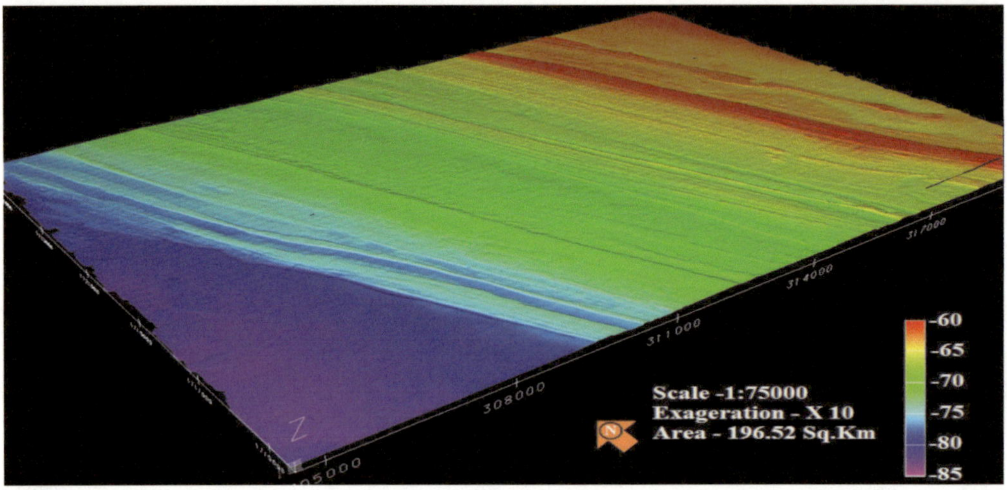

Fig. 3. Multi-beam bathymetry map showing small scale features off Goa covering distances along northern and southern axes are 17.50 km and 13.30 km respectively.

2.2 Bathymetry of deeper areas

In (Fig. 5) deep water areas from three Indian Ocean regions are presented. Locations of the present interest include: West of the Andaman Island (WAI), Western Continental Margin of India (WCMI) and Central Indian Ocean Basin (CIOB). The topographic data was collected using multi-beam echo sounder Hydrosweep system. Quantitative estimation of roughness parameters from a range of seafloor areas may help to understand the genetic linkages among the area seafloor features (Chakraborty et. al, 2007). Bathymetric data from the Andaman subduction zone in the Bay of Bengal (site A) consists of plain, trench, slope

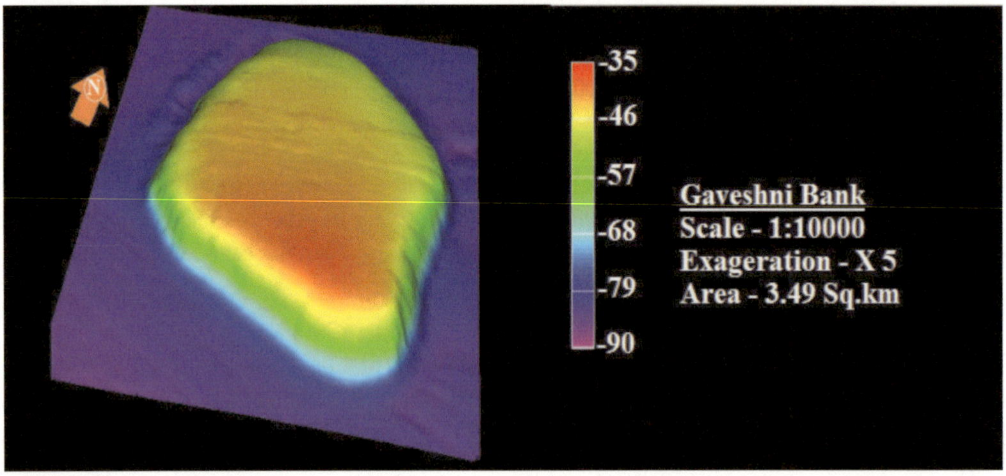

Fig. 4. Multi-beam map of the coral bank- Gaveshani bank covering distances along northern and southern axes are 2.40 km and 1.40 km respectively.

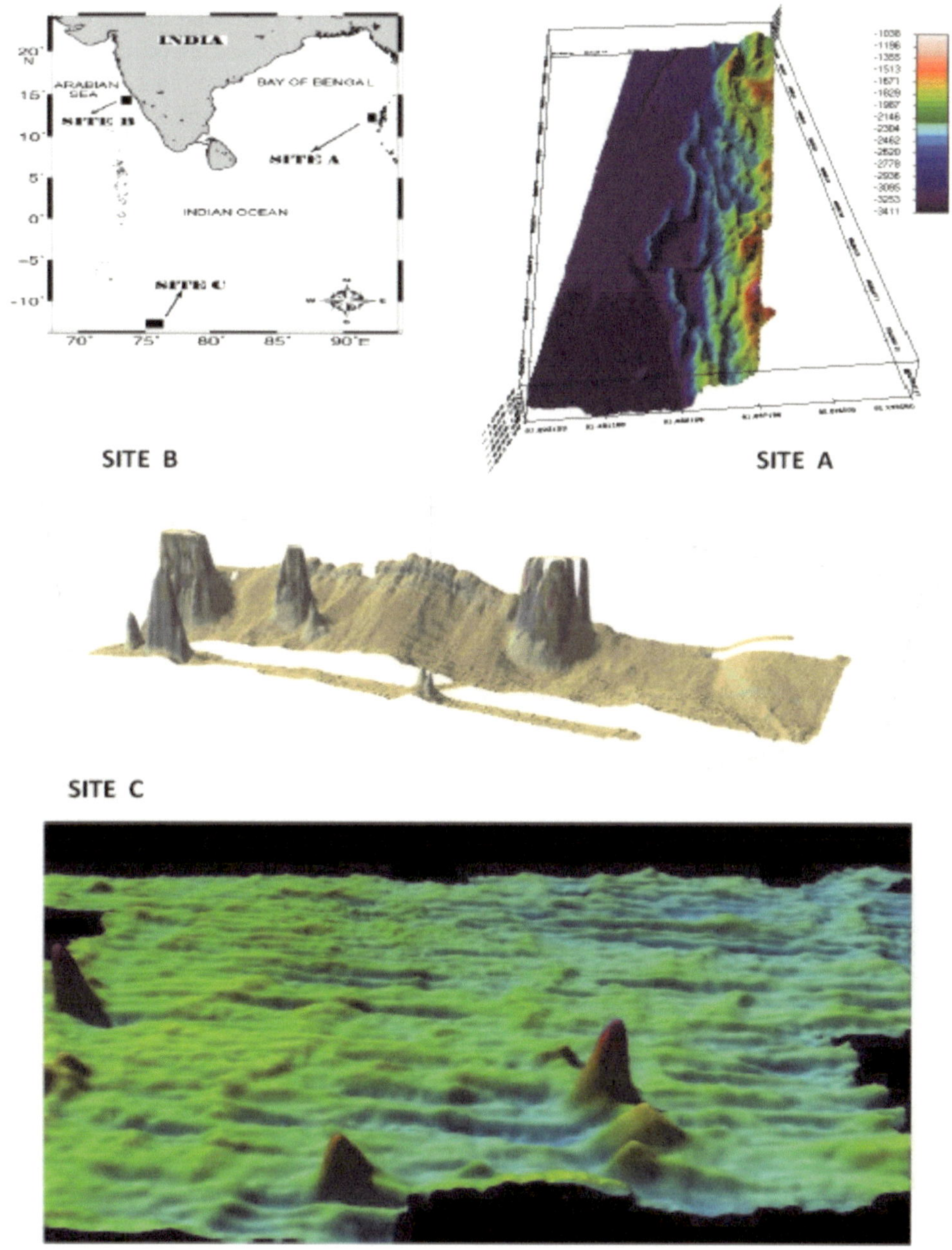

Fig. 5. Multi-beam Digital Terrain Model (DTM) & 3D perspective maps of the deep water sites; 'A'- (West of Andaman Island:WAI) (Chakraborty and Mukhopadhyay, 2006); 'B'- (Western Continental Margin of India : WCMI) (Chakraborty et al, 2006); SITE 'C'-(Central Indian Ocean Basin: CIOB) (unpublished data). Please see Fig. 6 for locations and dimensional details etc. of the study area.

and ridge off the Andaman Islands (Chakraborty and Mukhopadhyay, 2006). An area covering 25,000 km^2 around Andaman subduction zone was used to produce swath bathymetric map (Fig. 5). In this area the heavier Indian plate is shoving below the lighter continental Eurasian plate and causes the subduction morphology and associated structural features. The depth values were found to be varying between the 3600 m (western end of the trench side) to ridge areas (towards the Island side) having seafloor depth of 1000 m. In this region, the bathymetric grid cell size is taken to be 150 m, and the bathymetry map is presented. Data along three survey tracks (A to C) (Fig. 6) across the trench and other areas were collected to constrain exact physiographic profiles of the trench and the ridge. About 20,500 km^2 area was surveyed off the central west coast of India (WCMI) (Chakraborty et. al, 2006 & Mukhopadhyay et al, 2008). WCMI data (site B) was acquired between the Karwar and the Kasargod. As mentioned, the seafloor topography in the area varies from nearly flat continental shelf to lower slopes (Fig. 6). The slope morphology appears to have modified by the presence of physiographic highs of varied dimensions and slump related features. In site B (WCMI) depth varies from shallow water (~ 64 m) to deep water (~2200 m). In order to obtain evenly spaced data, the raw data were first gridded. In site B, 100 m was found to be a suitable grid cell size, later on three bathymetric profiles namely D, E, and F covering outer shelf, Mid-Shelf Basement Ridge upper slope, and Prathap ridge of the basin were extracted respectively. In CIOB (site C), the gridding was carried out using suitable grid cell size of 200 m. Bathymetric maps (Fig. 5) were prepared having water depth varying between 4200 - 5800 m and later on five bathymetric profiles were extracted G-K. Of these, three E-W profiles (G-I) are situated in the northern and southern end. Other two N-S profiles (J and K) are situated along west and eastern end. It has been observed that, the western region of the site C is comparatively shallower than the eastern side and the seafloor morphology varies from medium to large scale and has E-W trending in the central part of the area. A chain of seamounts trending along N-S direction was also observed in the extreme eastern region of the study area. Also, few seamounts were observed in the south-west section of the map. The five depth profiles (G, H, I, J and K) from grid data are studied for roughness information from north central, south, east, and west part of the survey area respectively.

3. Bathymetric technique using multi-beam

3.1 Interpreting bathymetric data to understand the seafloor

Measuring the seafloor roughness and associating it to different morphological processes is a major goal of this section.

3.1.1 Spectral parameter estimation

To estimate the spectral or power law parameter, Welch method (referred in Chakraborty et al., 2007) was used. In this method, all the input series or the bathymetric profile is first de-trended to remove the best fit linear trend of the original profile. Then the first and last 10% of each profile is tapered with a cos^2 function. This minimizes the edge effects. A Fast Fourier Transform (FFT) algorithm is then applied, and the resulting complex spectrum is squared to obtain a periodogram, which is an unbiased estimator of the true power. The periodogram is presented as a function of wave number k (cycles per kilometer), and the results are then plotted in a $\log_{10}$-$\log_{10}$ plot. This form of power spectrum was originally suggested (Malinverno, 1989) on the basis that the topographic profiles are self-affine and concluded that different depth profiles may be characterized by different fractal dimension (D). Using regression technique, the power law on a logarithmic scale can be written as: $\log_{10} P_H = (-\beta)$

$\log_{10} k + \log_{10} c$. Comparing this equation with the straight-line equation reveals that 'β' corresponds to the slope of a straight line fitted to the periodogram and 'c' corresponds to the intercept. At this stage the straight regression line fitting is done for the entire range of $\log_{10}$ (wave number) of the periodogram. The detailed post processing activity is shown (Fig. 7).

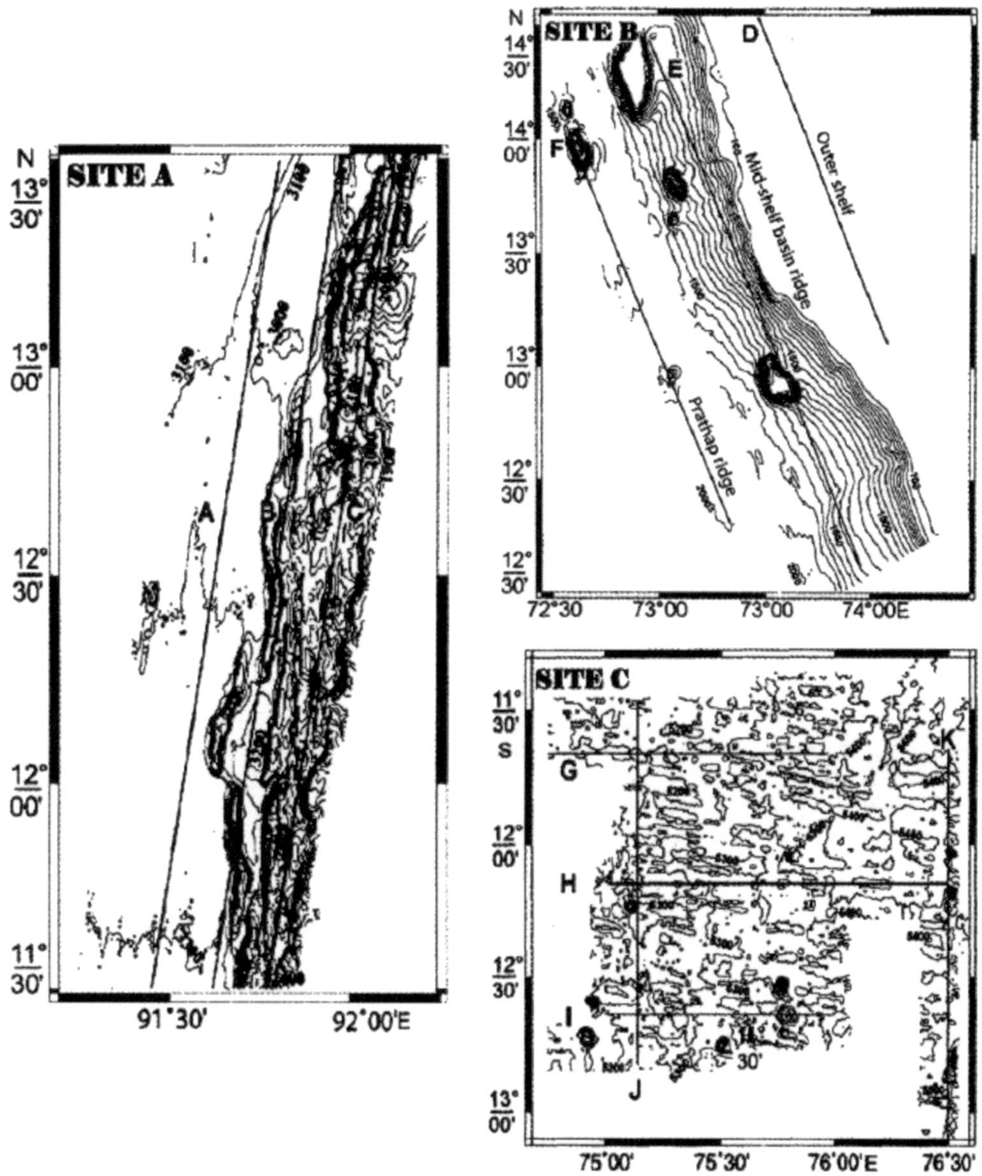

Fig. 6. Contour maps and chosen tracks for data analyses (adapted from Chakraborty et al., 2007). Sites; 'A'- (West of Andaman Island :WAI) (total area: 25000 km^2, maximum: 3600m and minimum depth: 1000m); 'B'- (Western Continental Margin of India : WCMI) (total area: 25500 km^2, maximum: 2200m and minimum depth: 64m); 'C' - (Central Indian Ocean Basin: CIOB) (total area: 22000 km^2, maximum: 5800m and minimum depth: 4200m).

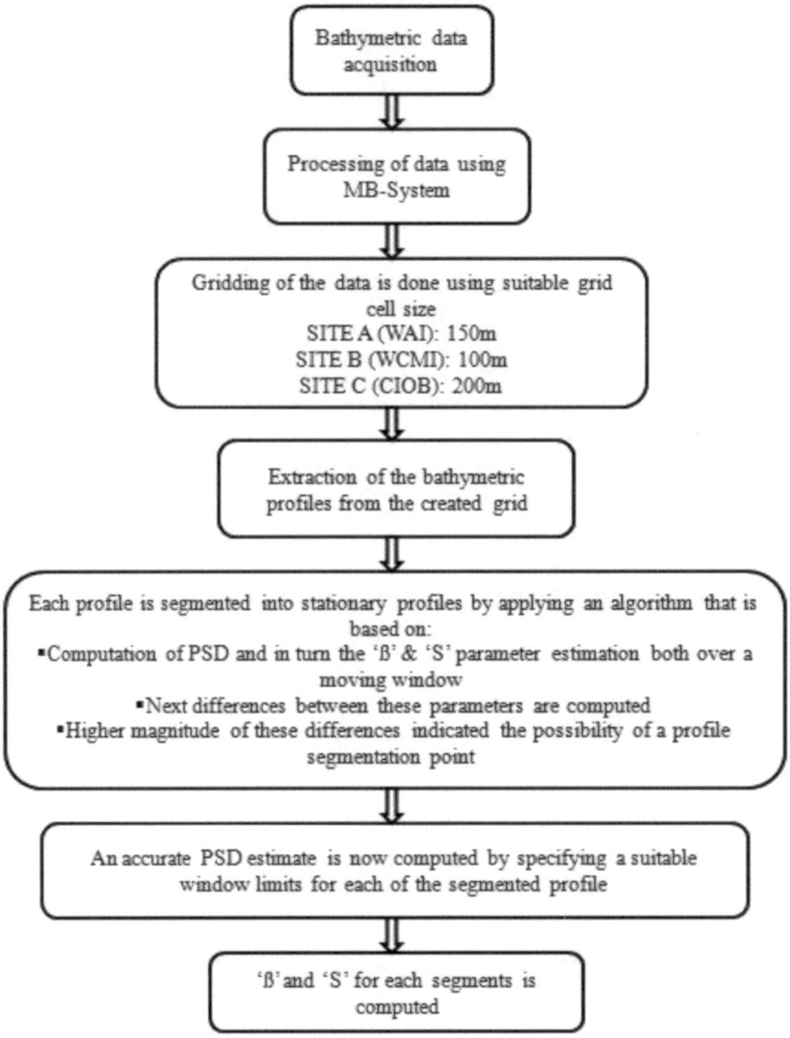

Fig. 7. Flow chart provides data processing and analyses steps (adapted from Chakraborty et al, 2007).

3.1.2 Amplitude parameter

The second parameter that is used to characterize the roughness of the profile is an amplitude parameter (S). A robust estimate is the median of the absolute deviation (MAD) from the sample median is given below:

$$MAD = \text{median} \mid z(x) - \text{MEDIAN}\,[\Delta z(x)] \mid \quad (1)$$

where $\Delta z(x)$ is the first differences of the depth series $z(x)$. An estimate of 'S' from MAD is defined as:

$$S = 1.4826 * MAD \quad (2)$$

Segmentation carried out to nine long profiles from three different physiographic provinces give 35 segments. Three profiles (A, B and C) for site A of WAI region provide 12 segments (Fig. 8). Our algorithm offers no segmentation to profile A from deeper section of this area having depth range higher than the 3000 m and 225 km of profile length towards the sedimentary provinces of trench side (Fig.5). Four segments were obtained for profile B which lies between the sedimentary trench and ridge area having an undulating seafloor at similar depth range. For profile C from relatively shallower depth section of ridge area provides seven segments at depth range 2000 -2400 m. Similarly, for site B from WCMI the profile D from shallower shelf (depth within the 100 m) presents only two segments. Eight topographic segments were obtained for profile E from the Mid-Shelf Basement Ridge. Distinctly different segments for flank parts (E4 and E6) along the bathymetric high E5 of the profile E are seen (Fig. 8). However, four segments obtained for deeper section of the profile are from sedimentary areas of the Prathap Ridge (conical features of the heights in Fig. 5). For this profile F, two highs having significantly different heights, are distinctly segmented. Overall, fourteen segments are made for this part (site B) of the physiographic provinces. For CIOB (site C), five profiles provide only nine segments from the deeper part of the seafloor (~ 5000 m depth) (Figure: 8). From this basinal part of the seafloor, west to east profiles G and H from the northern and middle part of the area do not offer any segment. Interestingly, remaining depth profile (I) from E-W direction provides four segments. Profiles J from western and K from eastern end provide two and one segments respectively. Once segmentation of the profiles are carried out, the estimation of power law exponent (β) for segmented profiles are important for calculation of fractal dimension. The bathymetric profiles extracted from the three different sites viz. A, B and C have sampling intervals of 150 m, 100 m and 200 m respectively. Hence, the extent of the periodogram on the higher wave number side in terms of $\log_{10}$(wave number) is limited to 0.50, 0.65 and 0.40 for site A, B and C respectively i.e., the true drop in power occurs somewhere between the wavelengths ~20 to ~0.2 km for Hydrosweep system operating at 15.5 kHz frequency (Fig. 8). Hence suitable limits are applied to the periodograms for curve fitting in order to estimate the spectral parameters (power law). In Fig. 8, power spectral density versus wave number plots for segmented sections (C4, E2, and J1) of the profiles C, E and J are presented respectively as an example.

3.1.3 Fine scale analyses using 'β' and 'S' parameters

Overall, the 'β' values are always negative and having large magnitudes, representing terrains which are relatively rough at large and smooth at smaller scales. Conversely, lower magnitudes of the 'β' parameter represent terrains which are smoother at large scales than at small scales. The scatter diagram between the power spectral exponent (β) and amplitude (S) parameters offer interesting results (Fig. 8). Overall variations in the roughness of these study areas may group in distinct clusters. For WAI physiographic provinces, the seafloor along the profile from sedimentary provinces of the trench side of site A possesses β and S values: –0.93 and 3.55 respectively. Profile A from sedimentary provinces of the trench side possesses β and S values: –0.93 and 3.55 respectively. As mentioned, high β corresponds to rough seafloor, while S quantifies amplitude (overall energy) of an area. For example, profiles having similar β but higher S will correspond to rough profile. Diagram (Fig. 8a) for this physiographic province indicates an isolated point (A1) for a rough seafloor from trench side. However, four segmented profiles (B1– B4) between the trench and ridge side provide (β= -1.87 and S= 6.50), (β= -1.64 and S= 10.20), (β= -2.00 and S= 8.86) and (β= -1.77 and S= 13.90) respectively. Similarly, for seven segmented sections of profile C from adjacent ridge area from Andaman Island provide variations of spectral parameter (β) value between (–2.19 to –3.02) and amplitude parameter S value (15.20 to 3.20). A critical analysis of these results indicate significantly lower β in comparison with the segments of the profile A i.e., smooth seafloor than the segments from the profile B & C. No clear cut cluster formation is seen in WAI region. Moderately to higher amplitude (S) parameters for profiles B&C data and relatively higher β value for profile A is indicative of higher large scale (B&C) and small scale (A) seafloor roughness respectively.

For site B of the WCMI, estimated 'β' and 'S' values of fourteen segmented sections are found to vary between (-3.56 and -0.90), and (0.24 and 22.28) respectively. The scatter diagram between the power spectral exponent (β) and amplitude (S) parameters offer interesting results (Fig. 8). The overall roughness observed in the study area groups in two distinct clusters. Cluster one includes segments from structural rises like: E6 (flank related to Mid-Shelf Basin Ridge), F1 (highs related to the Prathap Ridge) and F3 (S=22.28 consisting of two to Mid-Shelf Basin Ridge), F1 (highs related to the Prathap Ridge) and F3 (S=22.28 consisting of two small closely associated highs and not included in Fig. 8). These areas show high S and low to medium β values, which suggest large scale seafloor roughness. In contrast to this, another cluster comprises the remainder of the areas, including a gently sloping outer shelf and part of the western shelf margin basins. The seafloor here shows low S and moderate to higher β value, suggesting dominantly small-scale undulations. A comparison of the clustering of the structural rises with that of the outer continental shelf and basin regions (Fig. 8) indicate differences in their mode of origin (volcanic and plutonic, respectively). For example, the Mid-Shelf Basement Ridge and the Prathap Ridge are believed to have been formed by emplacement of volcanic material during the separation of India and Madagascar in mid-Cretaceous times. Detailed tectonic and other aspects of these areas in terms of clustering are well covered in Chakraborty *et al (2006)*.

The analyses based on the spectral and amplitude parameters in the study site C (CIOB) has a seafloor depth vary between 4200 m and 6000 m depth. The estimated 'β' and 'S' values of the E-W profiles (G, H and I) are found to be varying between (-1.14 and -2.70) and (7.44 and 4.87) respectively (Fig. 8). Also 'β' and 'S' values of N-S profiles are found to vary between (-1.70 and -2.083) and (13.33 and 12.75) respectively. If we remove data points related to the southernmost E-W segmented sections (I2, I3 and I4) of the profile I, the β parameter becomes comparable (-1.70 and -2.10) with N-S profiles (J and K), though, the amplitude parameter (S) are higher for N-S oriented profiles varying between (12.75 and 13.49) than the E-W profiles (4.9 and 7.9). Though we observe most of the profiles / segmented sections are similar, as far as β is concerned, however, show clear-cut existence of two clusters due to variation in amplitude parameter. The amplitude parameters show difference between the E-W and N-S profiles i.e., the amplitude (S) for N-S oriented profiles are higher than the E-W profiles (Kodagali & Sudhakar, 1993). For similar β values towards the E-W and N-S direction with higher amplitude for N-S profiles than the E-W indicate dominant large scale roughness for N-S profiles whereas E-W profiles indicate dominant small scale roughness. This fact is prominently clear in the report (see Fig. 5) (Malinverno, 1990).

3.2 Beam-forming technique to generate multi-beam data

In a multi-narrow beam system, geometrically a cross fan beam is created using two hull mounted linear/arc transducer arrays at right angles to each other. A narrow beam is produced in a direction which is perpendicular to the transducer's short axis. In general, the arrays are designed using a number of identical transducer elements which are equally spaced and driven individually. An assumption of an array which is kept at the origin of the co-ordinate system in a xy- plane, is made here. The separation between the elements is equivalent to 0.5 λ, where λ is assumed to design wavelength of the transducer array. The farfield pattern for a linear array is given by (Chakraborty, 1986):

The array factor f (θ, φ) is presented by as:

$$|F(\theta, \varphi)| = |f(\theta, \varphi)|\,|\text{Element pattern}| \tag{3}$$

$$f(\theta, \varphi) = \sum_{n=1}^{n} A_n \exp(j\,k\,r_m.\,\mathbf{R})$$

where A_n is the complex excitation coefficient, which is assumed to be unity and $k = 2\pi / \lambda$ and n is the number of array elements. For above linear array: $\mathbf{r_n} = d_n\,\mathbf{i}$, where d_n is the element separation from the origin, and **i** is the unit vector along the x-axis. The polar unit vector **R** can be expressed as:

$$\mathbf{R} = (\text{Sin}\,\theta\ \text{Cos}\,\varphi)\,\mathbf{i}$$

The array factor for a linear array can be written as:

$$f(\theta, \varphi) = \sum_{n=1}^{n} A_n \exp(j\,k\,d_n\,\text{Sin}\,\theta\ \text{Cos}\,\varphi)$$

The amount of phase delay required between the transmitting elements to steer the beam along the specified direction (θ_o, φ_o) may be expressed as:

$$\alpha_n = d_n.\,k\,(\text{Sin}\,\theta_0\ \text{Cos}\,\varphi_0)\,\mathbf{i}$$

The above term is to be applied to the far-field pattern of a uniformly excited linear array of the additive type. The farfield beam pattern in the final form is rewritten as:

$$f(\theta, \varphi) = \sum_{n=1}^{n} A_n \exp j\,\{(k\,d_n\,\text{Sin}\,\theta\ \text{Cos}\,\varphi) - (d_n.\,k\,\text{Sin}\,\theta_0\ \text{Cos}\,\varphi_0)\} \tag{4}$$

In equation (3)

$$\text{Element pattern} = \text{Sin}\,(\pi\,l\,\text{Sin}\,\theta\ \text{Sin}\,\varphi)\,/\,\pi\,(\text{Sin}\,\theta\ \text{Sin}\,\varphi) \tag{5}$$

where l is the length of the element and is equivalent to the 3.0 λ. The final expression for the vertical farfield pattern can be obtained by substituting equation (4) and equation (5) in equation (3). The farfield pattern of a multi-beam array of 40 elements is shown (Figure 9). The half power beam widths slowly increase: 3.0° to 3.6° as the beam steering angles increases. The computed half power beam-widths are generally narrow and are useful for high resolution bathymetric applications. Sidelobe 1evels around -13 dB for all the incidence angles are also seen. Sidelobes are also a concern in echo-sounding. There are certain techniques for suppression of the sidelobes, which significantly affect the performance of the array. For multibeam applications where many beams for different look directions are forming at a particular instant of time, sidelobes are assumed to be suppressed below -25 dB. This is necessary to avoid interference between the sidelobes and the main beam of the other directions. These sidelobe suppressions are accomplished by using Dolph-Chebychev window methods. Apart from that, sidelobe levels cannot be suppressed extensively beyond certain levels. Because it increases the width of the main beam and thus decreases the resolution. So suppressed sidelobe beam patterns around -25 dB for multibeam transducer array are acceptable as considered by echo-sounder designers (de Moustier, 1993).

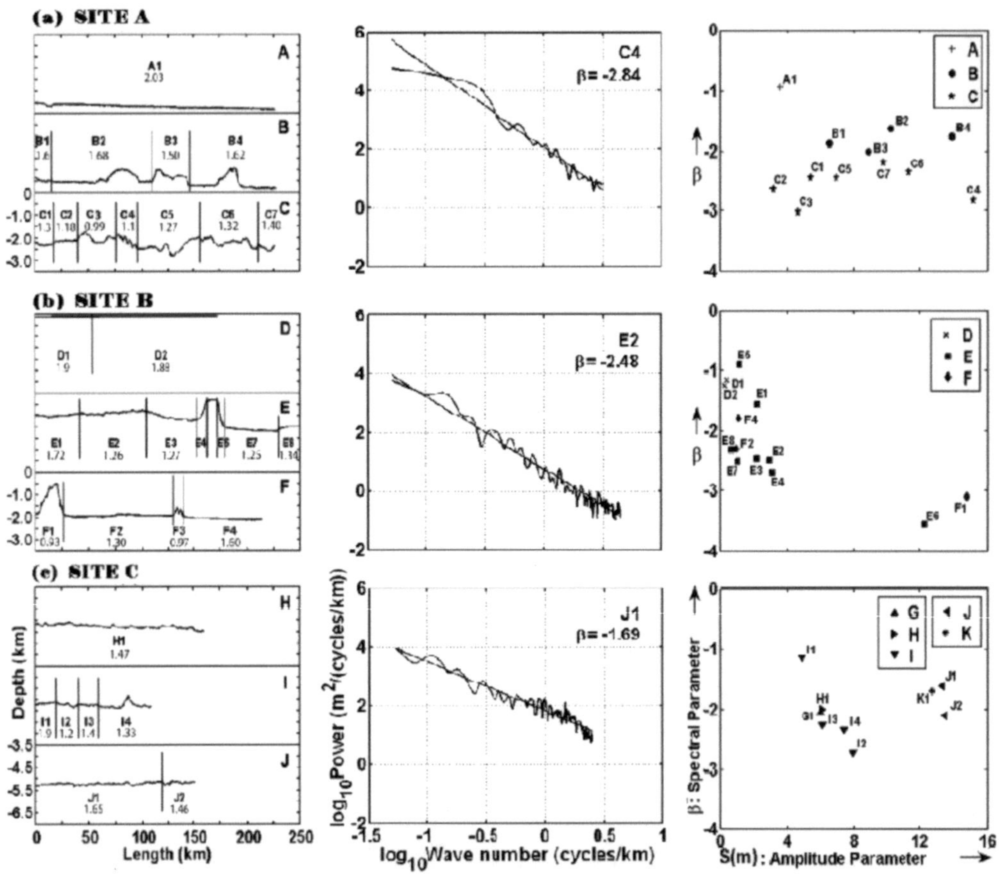

Fig. 8. Segmentation of the representative track lines and power law fitted straight lines and scatter plot of three geological provinces (adapted from Chakraborty et al., 2007).

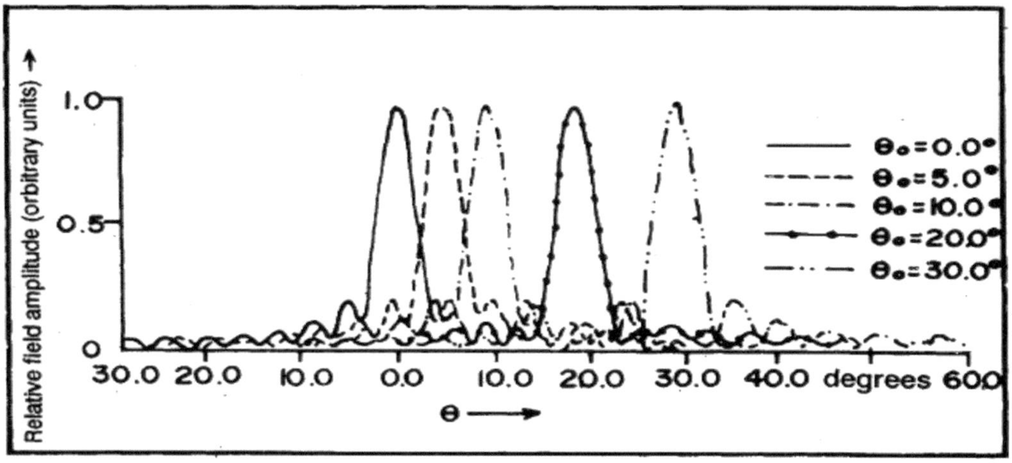

Fig. 9. Beam pattern for a 40- element linear array at different directions (adapted from Nair and Chakraborty, 1997).

3.2.1 Importance of sound velocity and other parameters for bathymetry

Using array transducers for beam-forming purposes the preformed beam directions are dependent on the used acoustic wavelength which is a function of the sound speed C, i.e $\lambda=C/f$, where C is in m/s and f is in Hz. For any changes in the sound velocity (C), λ will change. The sound speed is dependent on temperature, salinity and pressure (depth). Chen & Millero (1977) had proposed an expression to compute the sound velocity. Sound speed changes occur mostly due to temperature and salinity variations. Hence, variations of λ value with the variations of sound velocity, will change the half power beam-width of the array and also shift the beam direction. Sound speed changes of 6.6 percent are possible for an echo-sounding system operating at arctic and Tropical waters (temperature difference of 30°C). Similarly for λ variation of 3.3 percent, a beam rotation of 0 to 45° will introduce an error amount of 1.9° for a narrow beam of 2° beam-width. Also for swath bathymetry application, the variation of sound speed over the entire water column must be considered. The angle of arrival of the different beam is affected due to the refraction effect which is dominant, for the outer beams of the array system. In order to correct: the refraction error in echo-sounding systems, computation of the sound system should be performed by integrating the sound speed profile from the depth of the array to the bottom. The harmonic mean of the sound, speed is preferred over the average sound speed (Maul & Bishop, 1970 referred in Nair and Chakraborty, 1997). The multi-beam echo-sounder systems acquire bathymetry data along with other data can be used in scientific studies. As observed most of the multi-beam data is occasionally recorded with errors (de Moustier and Kleinrock, 1986), are either geometrical or refraction in nature. But to obtain a map that can be used either for navigation or scientific purposes, it is necessary that the data is error free. As per IHO standards the refraction affected data in shallow waters include uncertainties from navigation point of view. Also, to use in scientific applications it is vital to eliminate these errors from the bathymetry data. The refraction error influences the slant ranged beams and can be easily identified as the swath shape deviate in creating artificial features known as refraction artifacts.

For development of echo-sounding devices, certain parameters are of paramount importance to obtain optimum signal to noise ratio. Source level *i.e.* the transmitted power measured at 1m from the transmitting array is an important parameter. Similarly, transmission loss in the medium is also compensated at TVG (Time Varied Gain) module of the receiver. Apart from that, the noise level is also another important issue for the designer of the multibeam system. The noise level parameter is dictated by the location of the array in the hull. i.e., array should be kept in the forward part of the ship's hull, which is closest to the central line. The machinery noise is minimum in this area of the ship. The positioning of the array should also be decided based upon the ship's movement due to roll and pitch. Therefore, the array should be maintained submerged throughout the ship's movement. The major signal processing event which is known as beam-forming takes place at the receiver end of the array. The received signals are pre-amplified and each channel signal (array element) undergoes a correction for TVG loss in the correction unit and then beam forming techniques are applied to obtain multiple beams using predetermined delays for different directions. The theoretical backgrounds of the multi-beam signal processing methods have already been mentioned. The beam-formed waveforms are tracked at a bottom echo module for different preformed beam directions in order to determine the depth.

3.2.2 High resolution beam-forming

In multibeam sounding system, the delay-sum method (the method mentioned above) is used to obtain the beam-formed output in different directions. The size of the transducer array is a function of the beam width i.e., the spatial resolution of the system. At 15 kHz design frequency for a beam width of 2°, the approximate size of the transducer array has to be around 3 meters, which is relatively bulky. So a study is important to examine the effect of various methods of beam-forming which require a smaller array size by using increased signal processing alternatives to the dry end of the multibeam system. These techniques are the 'Maximum Likelihood Method' (MLM), and 'Maximum Entropy Method' (MEM) (Jantti, 1989 & Chakraborty &: Schenke, 1994).

The situation is more complicated when multibeam systems are operated over seamounts, slopes or ridge areas. The multiple sources of interference arrivals from different directions are well known. This condition significantly affects the map resolution so that an assumption of coherent and incoherent source arrival conditions is introduced to study the performance of the high resolution techniques. In Fig. 10, we present high resolution (MLM) beam patterns for incoherent and coherent sources. We assume that the sources are separated at 2° and 8° for different input signal to noise ratio. We observe that for closely placed sources (2°), the sources are unresolvable for coherent and incoherent sources. For large source separations of 8° the sources are resolvable for both the source conditions. The number of elements was chosen to be 16 which is one-fourth of the conventional array size used for multibeam systems. This techniques are affected due to the low signal to noise ratio conditions, and therefore needs to be studied using different algorithms such as ESPRIT, MUSIC, etc (Jantti, 1989). The high resolution beam-forming technique must also consider its use towards the real time application. It is unlikely that a single algorithm will satisfy all the necessary conditions to generate noise free bathymetry output. Development of a decision making network should be the next stage of development. This network will find suitable high resolution algorithms in real time and would be useful for multi-beam applications.

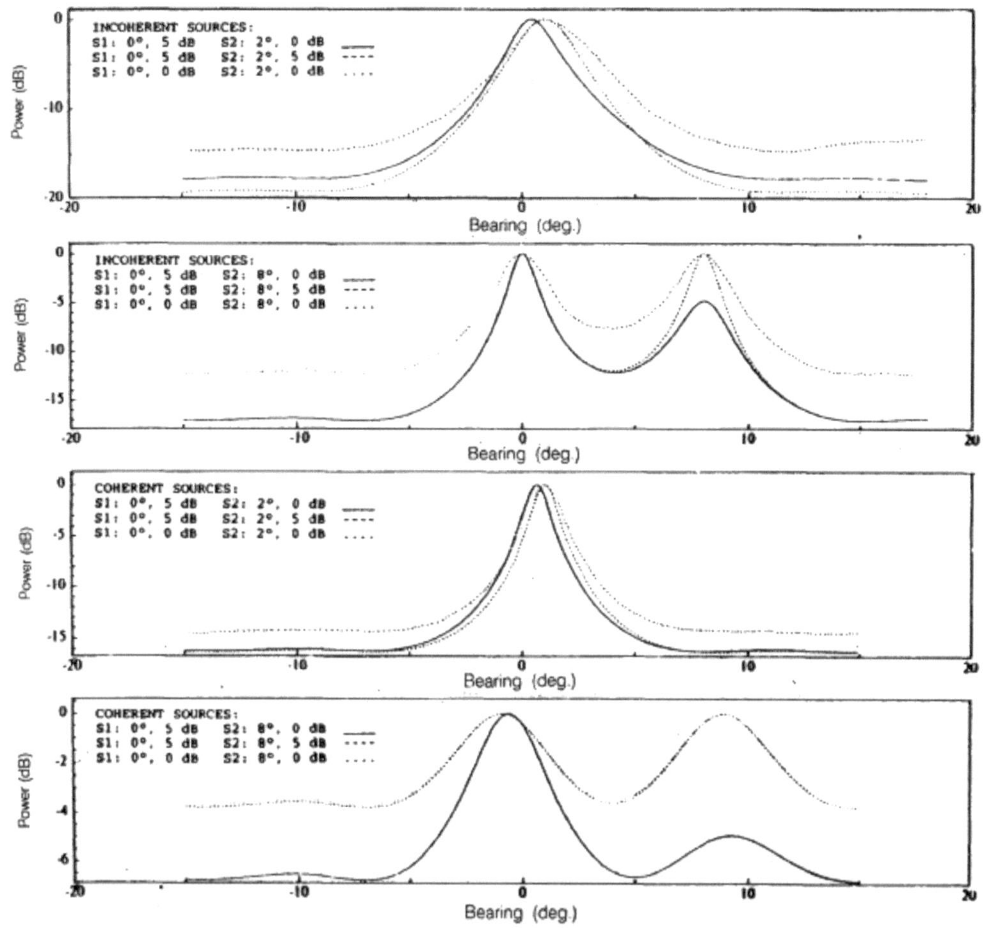

Fig. 10. Multi-beam High-Resolution beam patterns (MLM: Maximum Likelihood Method) for 16- element array. The sources are incoherent and coherent types of different signal strengths (adapted from Nair and Chakraborty, 1997).

4. Bathymetry and backscatter data

However, bathymetry only provides the shape of the seafloor features and depends on the resolution of the sounding system. To characterize the nature of the seafloor surveyed, studies of signal waveforms and their variations must be made, i.e. understanding of the backscatter signal is essential. Using the capabilities of multibeam system to provide backscatter signal from different incidence angle, modeling can be performed for seabed classification.

4.1 Angular backscatter data & quantitative seabed roughness

Multibeam systems are useful to derive angular dependent backscatter function of the seafloor. Besides bathymetry, different types of seafloor can be differentiated based on the backscatter values obtained at different incidence angle by the multibeam echo-sounder.

The aim of such study is to determine whether bottom types can be determined using angular backscatter functions. In addition, determination of the influence of the sea water-floor interface or sediment volume roughness parameter is possible. The Multibeam technique is particularly suitable to the task of deriving angular dependence of seafloor acoustic backscatter because it provides both high resolutions related to such measurements and bathymetry. Though it is difficult to calibrate the backscatter data, quantitative estimation may be made even for the relative values. The use of composite roughness theory (Jackson et al, 1986) is being made to obtain seafloor parameters. In the composite roughness theory, Helmholtz -Kirchhoff's interface scattering conditions and perturbation conditions were used including the volume scattering parameters. Extensive modeling studies have been carried out to use angular backscatter data to obtain quantitative roughness and sediment type parameters. A significant amount of work is carried out to obtain quantitative seafloor roughness parameters using bathymetric system such as multi-beam and single beam system (de Moustier & Alexandrou, 1991). The main challenge lies with this application is connected with the calibration, and work is still in progress to make a bathymetric system to obtain quantitative roughness parameters. An example of the use of inversion modeling from nodule bearing seafloor is provided (Fig. 11). The interface roughness and sediment volume parameters are estimated using the composite roughness theory (Chakraborty et al, 2003) from CIOB.

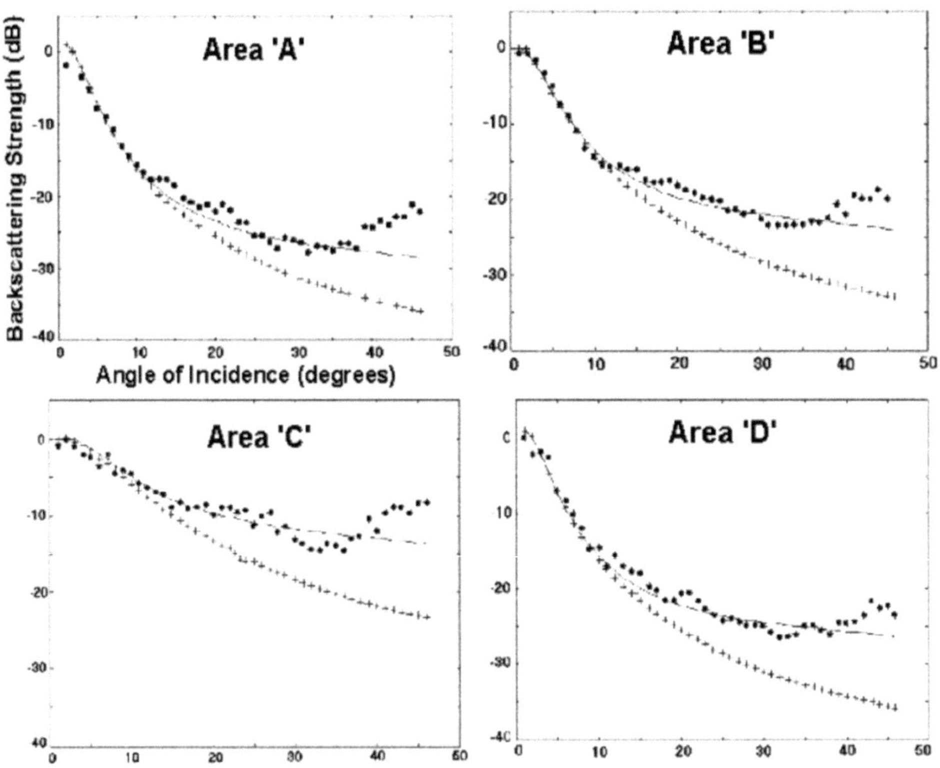

Fig. 11. Measured (dotted line in bold) and model backscatter strength (decibels) due to sea-floor-interface roughness (crossed line) and sediment volume roughness overlying sea-floor interface roughness (dashed line) versus incidence angle for four study areas (adapted from Chakraborty et al, 2003).

4.2 Seafloor backscatter image data

The side scan modifications to the multibeam system are also a popular technique. The raw side scan data requires significant image enhancements. Extensive work on image enhancements techniques has been carried out for side scan systems such as GLORIA and SEAMARC (Mitchell & Somers, 1994). Large scale implementation of these techniques is made for multibeam systems. Application of multi-beam bathymetric data provides interesting seafloor features. Besides obtaining bathymetric data for seabed characterization backscatter data is also important. However, use of raw multi-beam backscatter image data is limited due to the presence of inherent artifacts. Generally, angular backscatter strength data show higher values towards the normal incidence angles especially for smooth seafloor compared to outer-beam angles. Therefore, off-line corrections are essential to compensate outer-beam backscatter strength data in such a way that the effect of angular backscatter strength is removed. Moreover the effect of on line gain functions employed to the multi-beam is also to be removed apart from the effect of the large scale seafloor slope along and across track directions. These techniques employed to backscatter strengths data of the multi-beam system provides a normalized image of the seafloor suitable for studying sediment lithology.

EM 1002 multi-beam echo-sounding system installed onboard CRV *Sagar Sukti* has 111 pre-formed beams i.e., recording 111 depth values for a single ping. In addition to the depth, the system also provides quantitative seafloor-backscatter data, which is utilized to generate normalized backscatter imagery for the study of spatial distribution of fine scale seafloor roughness and textural parameter. Originally during the time of acquisition and consequent storage, individual angular backscatter strength data is corrected for number of losses and changes such as Time Varied Gain (TVG), i.e., propagation losses, predicted beam patterns and for the ensonified area (with the simplifying assumptions of a flat seafloor and Lambertian scattering). This data then gets recorded in a packet format called datagram stored for every ping. A very brief discussion of the developed algorithm named as *PROBASI (PROcessing BAckscatter SIgnal) II* applied to raw backscatter data of the EM 1002 multi-beam system in order to obtain normalized backscatter image data.

In EM 1002 system online amplifier gain correction is accomplished by applying mean backscattering coefficients such as BS_N and BS_O applied at 0° and at crossover incidence angles (normally 25°) respectively. Consequently the raw backscatter intensities recorded in the raw (*.all) files are corrected during data acquisition employing Lambert's law (Simrad Model). However, for lower incidence angles (within the $0\text{-}25^o$) the gain settings in the electronics require a reasonably smooth gain with incidence angle i.e., the gain between BS_N and BSo changes linearly. The sample amplitudes are also corrected suitably incorporating transmitted source level and transducer receiver sensitivity. Further, sonar image amplitudes, though corrected online, needs further improvement to generate normalized images for the seafloor area. This is especially needed for incidence beam angles within the $+/10^o$ angles (to remove routine artifacts in the raw backscatter data near normal incidence angles) (Fig. 12). Hence, post processing is essential to be carried out even for moderately rough seafloor. In addition to the artifacts close to normal incidence beams, the EM 1002 multi-beam data show some residual amplitude due to beam pattern effect, and thus real time system algorithm is unable to compensate such routine situations. As we know that, the EM1002 multi-beam echo-sounder system automatically carries out considerable amount of processing on raw backscatter intensities. Even then, the data show some residuals, which are required to be corrected before further studies. These fluctuations may be due to: Seabed

Angular Response and Transducer Beam Pattern Effects. Four basic corrections are made during the application of the PROBASI II software: data extraction and geometric corrections, heading correction, position correction, bathymetry slope correction and Lambert's law removal. In short the entire developmental activities are divided into four modules adopted through following procedure (Fernandes & Chakraborty 2009):

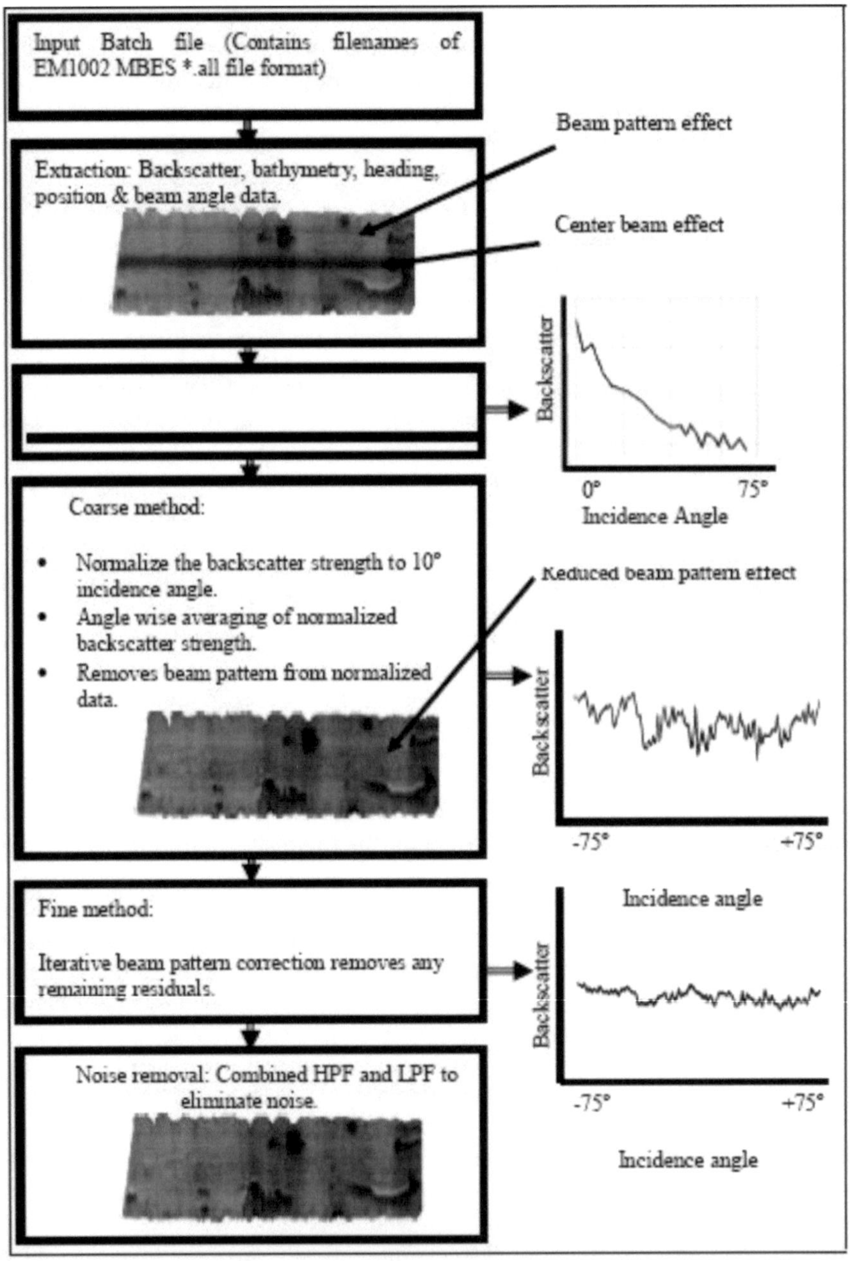

Fig. 12. Flow diagram of multi-beam processing algorithm (PROBASI II), and HPF & LPF are the high pass filter & low pass filter respectively.

Module-I: This extracts data from seabed image, raw range and beam angle, depth, attitude and position datagrams (EM Datagrams Manual). The module applies heading, position, slope correction on data and eliminates the effect of Simrad Model from data.

Module-II: It employs a coarse method. The module filters out maximum residuals or spikes in the backscatter strengths. This improves the image by 20% over the raw backscatter image.

Module-III: This module utilizes iterative beam pattern removal method in a finer sense. Since the coarse method does the maximum job of removing center beam and beam pattern effects. Still to some extent small beam pattern residuals remain in the data. The employed fine method removes such residuals more prominently.

Module-IV: The noise is an influencing factor on the quality of backscatter strengths. Since this type of data directly deals with the signal strength, sometimes the data gets recorded along with the noise occurring at the face of the transducer. Therefore, use of the filters becomes necessary. This module uses a combination of high pass and low pass filters.

4.3 Use of multi-beam bathymetry and backscatter data for pockmark studies

In this section we have presented a study example from western continental margin of India where backscatter image and bathymetric data is applied in a combined manner to generate a better picture of the proxies related to non-living resources such as pockmark (Dandapath et. al, 2010 and references therein). The pockmarks are potentially important for studies of marine resources and environments because of their relationship with venting of gas or other fluids generated by biogenic or thermogenic processes. Multibeam echo-sounders are potentially useful for locating and mapping these pockmarks. The presence of fluid escape features like seafloor pockmarks was first introduced by King & MacLean (1970). Generally, these are either hemispherical or disc shaped seafloor depressions having steep sides and flat floors. Very often, pockmarks exist in the continental margins (Hovland & Judd, 1988). In plan-view, these are usually circular, elliptical or elongated, and may be composite in shape. Present study area stretches over 105 km^2 offshore Goa along WCMI (Fig. 13), in water depths ranging from 145 m in the northeast to 330min the southwest. The average slope of the study area is 0.90^0, whereas the slope towards the shallower depth is 0.61^0, and towards the deeper side this slope changes over to 1.68^0. Rao et al., (1994) have reported that the recent clay-rich mud overlies the inner part of this slope area while relict sand is abundant in the outer slope area. In this work, using multibeam data, set of pockmarks were observed close to NNW-SSE trending fault zone. The margin is believed to have been formed in two phases in geological past (Mukhopadhyay et al., 2008). Interestingly, these pockmarks (total: 112) are observed nearly 50 km away from a BSR zone (Rao et al., 2003) marked in the Fig. (13b). Acoustic backscatter strength of the area ranges from -26 to -57 dB. Such backscatter variability is related to the seafloor slope, sediment type and relief (Blondel, 2009). Backscattering strengths at different water depth for each individual pockmark centre. In the deeper water (>210 m), many pockmarks show high backscatters centrally (-27 to -40 dB). Towards the west in deeper water, the seafloor has strong backscatter (-35 dB) suggesting coarser grained sediment at the seafloor because of increased acoustic impedance, and roughness-related scattering from coarse sediment. In shallow water (<210 m) where seabed gradients are gentle, normally backscatter strength is low. The

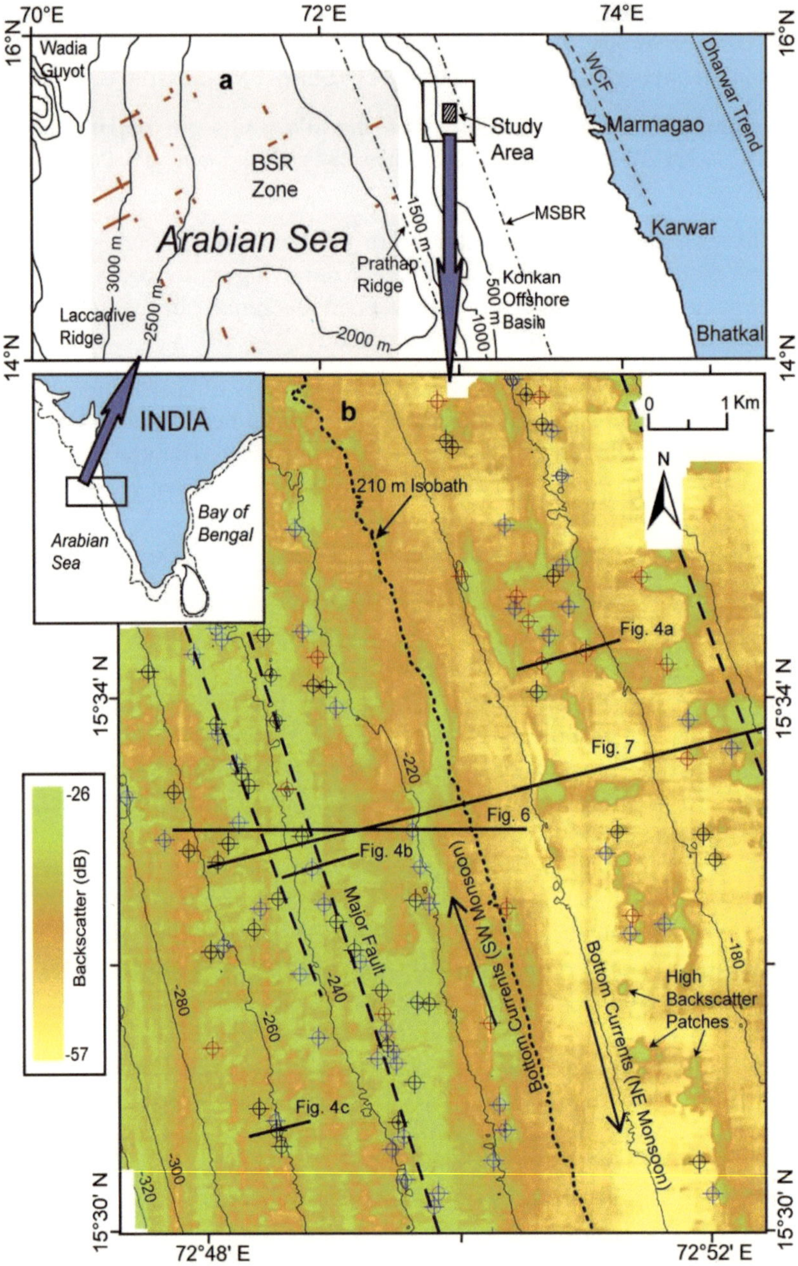

Fig. 13. (a) Location of study area. Red highlighted lines and grey shade indicate the identified bottom simulating reflectors. MSBR refers to Mid-Shelf Basement Ridge, and WCF indicate West Coast Fault. (b) Backscatter map of the study area showing isobaths with an interval of 20 m depth. Pockmarks are indicated by crossed circles. Black, blue and red circles with cross marking represent circular, elliptical and elongated pockmarks, respectively. Dashed lines indicate location of identified faults. Solid black lines represent location of the corresponding profiles for Fig 14. (adapted from Dandapath et al, 2010).

area is covered by soft terrigenous clayey mud producing average seafloor backscatter strength (-45 dB). Occasional discrete curvilinear, circular or clustered patches of higher backscatter (-32 dB) occur towards the shallower part (Fig. 14). Strong backscatter (-30 to -38 dB) is also observed around the fault. The variation in backscatter observed between these different areas is broadly comparable with results of calibrated acoustical measurements between different known sediment types.

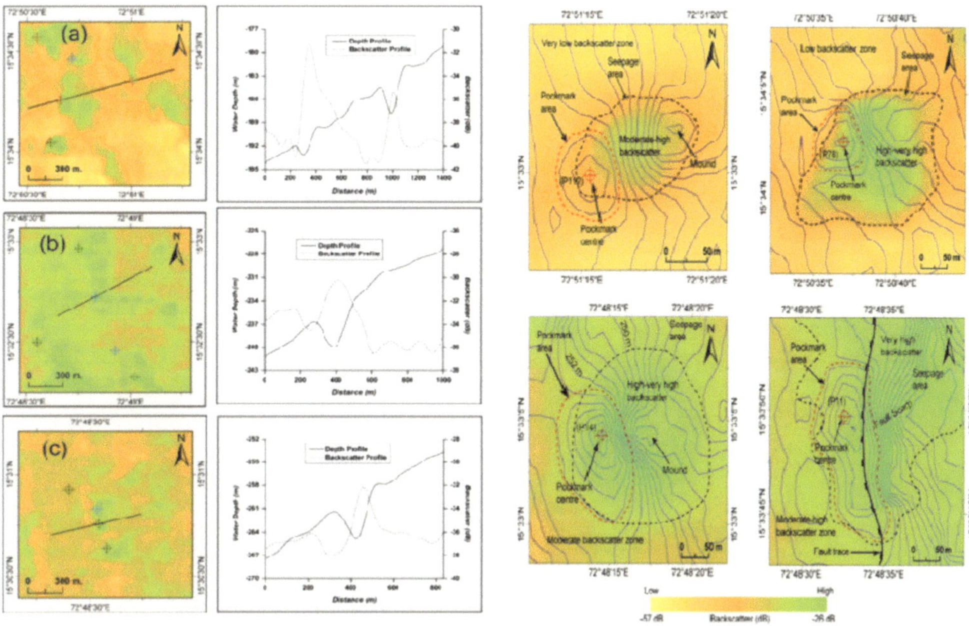

Fig. 14. Backscatter strength and bathymetric profiles for three locations and adjacent areas are shown. Stronger backscatter strength from the centres and upper sidewalls are distinctly indicated. Locations of these profiles are also marked in Fig. 13. (adapted from Dandapath et al., 2010)

We have visualized an abrupt increase in backscatter strength around the pockmark depressions (Fig. 14), and backscattering strength is also significantly higher within the pockmark-dominated areas, as reported elsewhere. The variability of backscatter strength of the pockmarks is also affected by slope, sediment type and relief of the seafloor. Average backscatter strength in the deeper area (-35 dB) is higher compared to the shallower area (-45 dB) due to different acoustic impedance and roughness-related scattering. High seafloor backscatter in the deeper pockmark zone is attributed to coarse sediment, which might have been left inside the pockmarks due to winnowing of fine grained sediments. Likewise, the coarse fraction could also result from the precipitation of diagenitic or authigenic minerals associated with fluid venting. Low backscatter strengths in the shallower areas are caused by different types of recent sediments (clays). In such areas, however, occasional high backscatter patches surrounded by low backscatter ones are associated with sediment movement as observed elsewhere. Although, the backscatter over the whole study area widely varies between (-26 to -57 dB), but within the pockmark itself, it is limited (-27 to -48 dB) i.e., much higher. In this section we have presented pockmark morphological parameters using multibeam bathymetry data and GIS. Combined observation of bathymetry and backscatter data enabled us to assess and estimate the morphological parameters of the pockmarks. We have detected a total of 112 pockmarks of which 43 are circular, 51 are elliptical and the remaining 18 are of elongated type (Fig. 15). A brief account of dimensions, shape, cross-section, orientations, and spatial distribution of the pockmarks is given (Dandapath et al, (2010) and references therein. Most pockmarks are small to medium sized, with lengths varying from 70m to 514m and widths from 50m to 136m. The average length and width are 157 m and 83 m, respectively. Pockmark vertical relief varies from 0.7 m to 5.0 m with an average of 1.9 m. In addition, there are 11 and 16 pockmarks of relief >3 m and <1 m respectively.

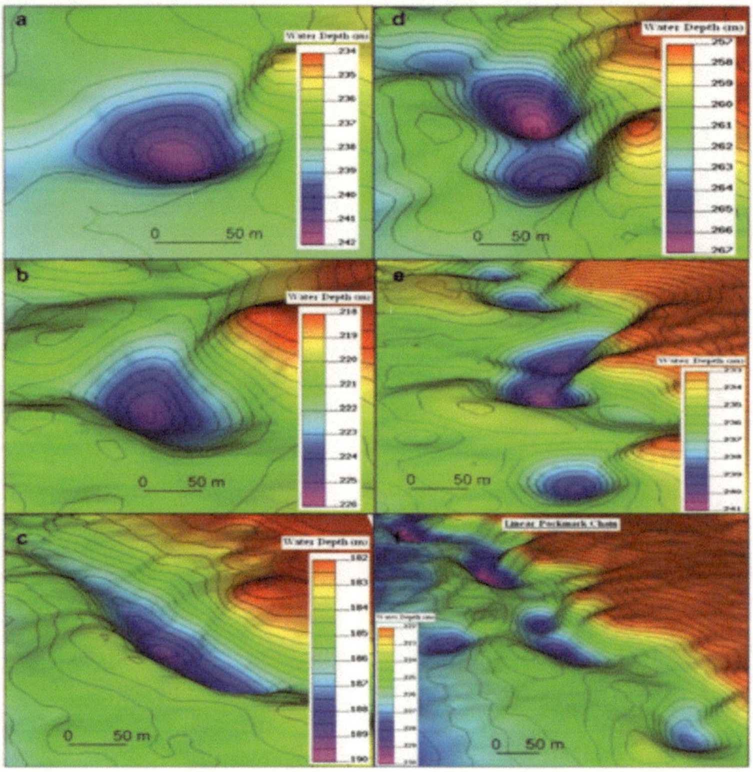

Fig. 15. Perspective view of bathymetry of typical pockmarks that are (a) circular, (b) elliptical, (c) elongated, (d) composite and (e) and (f) forming chain (scale is approximate; contour interval 0.5 m) (adapted from Dandapath et al., 2010).

5. Conclusions

The work embodied in this article presents the detailed methods to study the important aspects of the bathymetric techniques. The historical background involved in the development of the bathymetric system is outlined. Present day bathymetric systems such as single and multi-beam echo-sounding system developments have direct impact due to current technology level. For example: the wet end of the system i.e., transducer is developed significantly due to the material development taken place in last two decades. Moreover, development of computer technology allows faster beam-forming and associated signal processing techniques to work in real time. The involvement of GIS (Geographical Information System) in bathymetric mapping also has contributed superbly.

We have presented shallow water multi-beam bathymetric data from three geological important areas of the western continental shelf of India. The acquired data, and subsequent processed 3D output from three different environments visualizes the level of the technology to understand the earth related subject. The qualitative description of the seafloor processes along with the analyses presented here from "Fifty-Fathom Height" off Mumbai, northern part of Goa, and coral bank off Malpe generate significant interest. The numerical technique such as spectral methods employed to the bathymetric data of the deeper part of the western continental margins of India such as: western Andaman Island (WAI), western continental margins of India (WCMI), and central Indian Ocean basin (CIOB) provide power law parameters. The segmentation of the bathymetric data into linear form and subsequent estimation of the power law parameters through straight line fittings techniques have open up a method to analyze the seafloor roughness.

Bathymetry only provides depth data based on the functioning of the bottom tracking gate for echo-sounding system. The echo-waveform analyses to determine arrival time of the bottom echo depends on the echo-peak (near nadir beams) or energy values (off nadir beams). The analyses made on such data for roughness studies have limitations. Comparatively, use of seafloor backscatter data involves entire seafloor insonified area. Though, large scale roughness i.e., structural aspects can be estimated using bathymetry, but, in order to estimate roughness towards the small scale end (cm scale) the power law curves are needed to be extended over high-frequency end. The backscatter data are found to be useful when applied to study micro-topographic studies employing Jackson model (1986). Here, we have adopted similar techniques for multi-beam Hydrosweep system at CIOB.

Moreover, this work also presented backscatter image processing from pockmark dominated areas of the western continental margin of India. The technique associates bathymetry as well as backscatter data to provide pockmark morphology along with seepage details of the area as an interesting example for non-living resource related studies. However, no claim for completeness is being made in the present work. Nevertheless many issues of bathymetric systems are made, yielding a scope for future work on various aspects of the bathymetric systems. Certain issues such as bathymetry using phase measuring system (de Moustier, 1993) is not covered here, though, modern multi-beam systems possess hybrid techniques such as beam-forming as well as phase measuring system. The shallow water data presented throughout in this work have been acquired using such system e,g. EM 1002 multibeam system.

6. Acknowledgment

We acknowledge the Director, NIO for his permission to publish this work, and are thankful to Ryan Rundle, visiting student (Master degree) fellow under RISE research scheme and

Andrew Menezes for reading this manuscript. We also acknowledge the review of Dr. P. Blondel. We are thankful to the Ministry of Earth Sciences, New Delhi (Government of India) for their support. This is National Institute of Oceanography contribution 5066.

7. References

Blondel, P. (2009). *Handbook of Sidescan Sonar*, Springer/Praxis, Chichester, UK.

Chen, C.T. & Millero, F.J. (1977). Speed of sound in seawater in high pressure, *J. Acoust. Soc. Am.*, Vol.62, No. 5, pp. 1129-1135.

Chakraborty, B. (1986). Coaxial circular array: Study of farfield pattern and field frequency responses. *J. Acoust. Soc. Am.*, Vol.79, No. 4, pp. 1161-1163.

Chakraborty, B. & Schenke, H.W. (1995). Arc arrays: studies of high resolution techniques for multibeam bathymetric applications. *Ultrasonics*, Vol. 33, No. 6, pp. 457-461.

Chakraborty, B., Kodagali, V.N. & Baracho, J. (2003). Sea-floor classification using multibeam echo-sounding angular backscatter data: A real-time approach employing hybrid neural network architecture. *IEEE JOE*, Vol. 28, No. 1, pp. 121-128.

Chakraborty, B., Mukhopadhyay, R., Jauhari, P., Mahale, V.P., Shashikumar, K. & Rajesh, M. (2006). Fine-scale analysis of shelf slope physiography across the western continental margin of India. *Geo-Mar. Letter*, Vol. 26, No. 2, pp. 114-119.

Chakraborty, B. & Mukhopadhyay, R. (2006). Imaging trench-line disruptions: swath mapping of subduction zone. *Currrent Science*, Vol. 90, No. 10, (2006), pp. 1418-1421.

Chakraborty, B., Mahale, V.P., Shashikumar, K. & Srinivas, K. (2007). Quantitative characteristics of the Indian Ocean seafloor relief using fractal dimension. *Indian Jour. Mar. Sci. (Special Issue on Fractals in Marine Sciences)*, Vol. 36, No. 2, pp. 152-161.

Dandapath, S., Chakraborty, B., Karisiddaiah, S.M., Menezes, A.A.A., Ranade, G., Fernandes, W.A., Naik, D.K. & Prudhvi Raju, K.N. (2010). Morphology of pockmarks along the western continental margin of India: employing multibeam bathymetry and backscatter data. *Mar. Pet. Geol.*, Vol. 27, No.10, pp. 2107-2117.

de Moustier, C. & Kleinrock M. C. (1986). Bathymetry artifacts in the Sea Beam data: How to recognize them and what causes them. *J. Geophys. Res.* Vol. 91, No. B3, pp. 3407-3424.

de Moustier, C. & Alexandrou, D. (1991). Angular dependence of 12 kHz seafloor acoustic backscatter. *J. Acoust. Soc. Am.*, Vol.90, No. 1, pp. 522-531.

de Moustier, C. (1993). Signal processing for swath bathymetry and concurrent seafloor acoustic imaging, In: *Acoustic Signal Processing Ocean Exploration*, J.M.F. Moura and I.M.G. Lourtie (Eds), pp. 329-354.

Jackson, D.R., Winebrenner, D.P. & Ishimaru, A. (1986). Application of the composite roughness model to the high frequency bottom backscattering. *J. Acoust. Soc. Am.*, Vol. 79, No. 5, pp. 1410-1422.

Jantti, T.P. (1989). Trials and Experimental results of the ECHOES XD Multi-beam Echo sounder. *IEEE JOE*, Vol.14, No.4, pp. 306-313.

Fairbanks, R.G. (1989). A 17,000 year glacio-eustatic sea level record: influence of glacial melting rates on the Younger Dryas event and deep -ocean circulation. *Nature,* Vol. 342,(07 December, 1989), pp. 637-642.

Fernandes, W.A. & Chakraborty, B. (2009). Multi-beam backscatter image data processing techniques employed to EM 1002 system, *Proceedings of International Symposium on Ocean Electronics (SYMPOL-2009)*, Kochi; India; 18-20 Nov. 2009. pp. 93-99.

Hovland, M. & Judd, A.G. (1988). *Seabed pockmarks and seepages-Impact on Geology, Biology, and the Marine Environment,* Graham & Trotman, London, UK.

King, L.H. & MacLean, B. (1970). Pockmarks on the Scotian shelf. *Geol. Soc. Am. Bull.*, Vol.81, No.10, pp. 3141-3148.

Kodagali, V.N. & Sudhakar, M. (1993). Manganese nodule distribution in different topographic domains of the central Indian Basin. *Mar. Georesour. Geotechnol.* Vol.11, No. 4, pp., 293-309.

Lurton, X. (2002). *An Introduction to Underwater Acoustics Principles and Applications,* Springer/Praxis, Chichester, UK

Malinverno, A. (1989). Segmentation of topographic profiles of the seafloor based on the Self-Affine model, *IEEE JOE,* Vol. 14, No. 4, pp. 348-359.

Malinverno, A. (1990). A simple method to estimate the fractal dimension of a Self-Affine series. Geophys. Res. Letters, Vol. 17, No. 11, pp.1953-1956.

Mitchell, N.C. & Somers, M.L. (1989). Quantitative backscatter measurements with a long range Side-Scan Sonar. *IEEE JOE.* Vol. 14, No.4, pp. 368-374.

Mukhopadhyay, R., Rajesh, M., De, Sutirtha, Chakraborty, B. & Jauhari, P. (2008). Structural highs on the western continental slope of India: Implications for regional tectonics. *Geomorphology.* Vol.96, No. 1-2, pp. 48-61.

Nair, R. R. (1975). Nature & origin of small scale topographic prominences on western continental shelf of India. *Indian Jour. Mar. Sciences,* Vol. 4, pp. 25-29.

Nair, R.R. & Chakraborty, B. (1997). Study of multi-beam techniques for bathymetry and sea-bottom backscatter applications. *Jour. Mar. Atmos. Res.,* Vol. 1, No. 1, pp. 17-24.

Quasim, S. Z. & Nair, R. R. (1978). Occurrences of a bank with living corals off the south-west coast of India. *Indian Jour. Mar. Sciences,* Vol. 7, pp. 55-58.

Rao, V.P.,Veerayya, M., Nair, R.R., Dupeuble, P.A. & Lamboy, M. (1994). Late Quaternary Halimeda bioherms and aragonitic faecal pellet-dominated sediments on the carbonate platform of the western continental shelf of India. *Mar. Geol.* Vol.121, No.3-4, pp. 293-315.

Rao, V.P. & Wagle, B.G. (1997). Geomorphology and surficial geology of the western continental shelf and slope of India: A review. *Current Science.* Vol.73, No.4, pp. 330-350.

Rao, V.P., Montaggioni, L., Vora, K.H., Almeida, F., Rao, K.M. & Rajagopalan, G. (2003). Significance of relic carbonate deposits along the central and south-western of India for late Quatenary environmental and sea level changes. *Sediment. Geol.* Vol.159, No. 1-2, pp. 95-111.

Seibold, E, & Berger, W.H. (1993). *The seafloor: An introduction to marine Geology,* Springer-Verlag, Berlin, Germany.

2

Airborne Electromagnetic Bathymetry

Julian Vrbancich
Defence Science and Technology Organisation (DSTO)
Australia

1. Introduction

Traditional methods for measuring the water depth rely on sonar soundings. However airborne techniques offer the advantages of increased survey speed and operation over dangerous waters that may be affected by very strong tides and the presence of shoals and reefs that limit the operation of surface vessels. The airborne lidar method has been used very successfully for mapping coastal waters and relies on the difference in the time between a laser pulse (infra red wavelength) reflected from the sea surface and a separate laser pulse (blue-green wavelength) reflected from the sea floor. The depth of investigation, typically 50 to 70 m in ideal conditions, depends strongly on water clarity (turbidity), as well as other factors (e.g. sea state, surf zone, sea bottom reflection) and weather conditions. These lidar systems provide dense depth soundings, typically a grid with a laser spot spacing of 4 to 5 m, and they meet the accuracy standards of the International Hydrographic Organisation (IHO).

The airborne electromagnetic (AEM) bathymetry method is based on the AEM technique (Spies et al., 1998; Palacky & West, 1991) developed for geological exploration of electrically conducting targets, initially applied to mineral exploration. Since then, the AEM technique has also been applied to environmental studies, e.g. mapping hydrogeological features in alluvial aquifers (Dickinson et al., 2010); and salinity distribution (Fitterman and Deszcz-Pan, 1998; Hatch et al., 2010; Kirkegaard et al., 2011). The AEM method uses an airborne transmitter loop with a known magnetic moment (i.e. a magnetic dipole source) that generates a *primary* magnetic field to induce electrical currents in the ground. These currents establish a *secondary* magnetic field (the EM response) which is detected by the airborne receiver loop, shown schematically in Figure 1. The mathematical theory of this EM induction process is thoroughly reviewed by Ward & Hohman (1987), Wait (1982), Weaver (1994), West & Macnae (1991) and Grant & West (1965). The EM induction process causes the electromagnetic fields to *diffuse* slowly into the conductive medium. Nabighian (1979) showed that the transient time-domain electromagnetic (TEM) response (caused by a rapid turn-off of the transmitter current) "observed over a conducting half-space[1] or a layered earth can be represented by a simple current filament of the same shape as the transmitter loop moving downward and outwards (in the isotropic conductive medium) with a decreasing velocity and diminishing amplitude, resembling a system of "smoke rings"

[1] A "half-space" refers to an infinitely deep homogeneous earth (ground) with a given electrical conductivity.

blown by the transmitter loop" (Nabighian, 1979). This concept was later extended to a frequency domain EM response (Reid & Macnae, 1998) where the transmitter loop is powered by a continuous sinusoidal current at a pre-determined frequency.

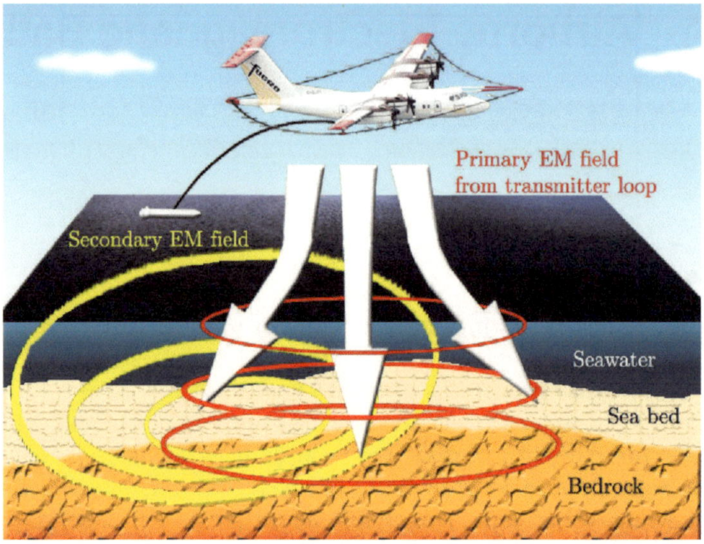

Fig. 1. Schematic diagram of the fixed-wing time-domain GEOTEM (Fugro Airborne Surveys Pty Ltd.) AEM system. The transmitter loop is mounted around the aircraft and the receiver loop is trailed behind and below the aircraft.

Thus the footprint associated with the EM induction process (to be discussed in section 3.4) is much larger than the lidar footprint. The physics of the diffusion of the EM fields in the conductive medium is such that for airborne (or ground) EM systems the horizontal resolution will always be larger than the horizontal resolution of lidar-based bathymetry systems because of the relatively small footprint of the spot laser beam. However, a distinct advantage of the AEM technique is that the EM response depends on the bulk conductivity of the medium (as well as other parameters), which in the case of seawater, is not affected by the turbidity or the presence of bubbles. Thus the interpretation of AEM data acquired over seawater is expected to provide water depths, in relatively shallow coastal waters irrespective of turbidity levels (within reasonable limits), and in the surf zone.

The simplest model for interpreting AEM data for bathymetry studies assumes a one-dimensional (1D) layered-earth structure which in its crudest form consists of two upper layers, a seawater layer overlying an unconsolidated sediment layer, with each layer defined by its electrical conductivity and thickness; these layers in turn overlie a relatively resistive[2] basement (bedrock). This two-layer model allows for a sediment layer (which may become vanishingly thin over exposed bedrock on the seafloor) and assumes no stratification in the conductivity of the seawater and sediment layers. Distinct conductive layers of seawater and/or sediment can be readily included by introducing more layers if warranted. The

[2] The electrical conductivity (S/m) is the reciprocal of electrical resistivity (Ωm) and these terms are interchanged accordingly, depending on whether one is discussing a good conductor or a poor conductor; e.g., seawater is conductive, bedrock is relatively resistive.

parameters used for inversion of the AEM data are the two thicknesses of the upper two layers, the three electrical conductivities associated with the two layers and the basement, and the elevations (and separations) of the transmitter and receiver loops above the sea surface. The thickness of the upper layer, as determined from the inversion of the AEM data, is the water depth which is the objective of the AEM bathymetry method. Where some parameters are measured directly, for example seawater conductivity and altimetry, or assumed, for example, sediment conductivity and basement resistivity, then these parameters may be held fixed or tightly constrained in the inversion process.

Given that the sediment layer and its associated conductivity are part of the layered-earth model, the thickness and conductivity of the sediment layer may also be determined in shallow waters from the AEM data. Combined with interpreted water depths (i.e. AEM bathymetry), the AEM method can therefore be used to (i) estimate the bedrock topography by combining the water depth and sediment thickness (Vrbancich, 2009; Vrbancich & Fullagar, 2007a) and (ii) map the seafloor resistivity (Won & Smits, 1986a). Combined with empirical relationships such as Archie's Equation (Archie, 1942) and assumed cementation factors for unconsolidated marine sediments (Glover, 2009), the derived seafloor resistivities can be used to estimate important seafloor properties such as density, porosity, sound speed and from these properties, acoustic reflectivity (Won & Smits, 1986a). The interpretation of AEM data to determine sediment thickness and sediment conductivity requires a well calibrated AEM system. The EM response is far more sensitive to the overlying conductive seawater layer than it is to the less conductive sediment layer and it is therefore more difficult to accurately determine the depth of the sediment-basement interface than the depth of the seawater-sediment interface. The same AEM dataset may provide bathymetric accuracy to within 1 metre and yet provide poor agreement with bedrock depths estimated from marine seismic data. The use of the AEM method for simultaneously mapping water depth and sediment thickness and conductivity is currently under investigation and for this purpose, it is important to have independent ground truth data consisting of (i) marine seismic survey data (matching the AEM survey line locations if possible) to provide an estimate of the depth to bedrock, and (ii) sediment resistivity data obtained, ideally, from bore hole sediment samples, or surficial sediment samples (extending to 3-5 m depth) obtained from vibrocore samples. A marine seismic survey and a resistivity study of vibrocore samples of shallow marine sediments has been undertaken in Jervis Bay (Vrbancich et al., 2011a) and Broken Bay (Vrbancich et al., 2011b), located approximately 150 km south and 40 km north of Sydney Harbour (~ 33.8° S, 151.3° E) respectively, to support the interpretation of AEM survey data to estimate depth of bedrock and sediment resistivity (Vrbancich, 2010).

1.1 Chapter outline

Section 2 discusses the initial development of AEM for bathymetric studies, and the associated application of AEM for sub-ice bathymetry and sea-ice thickness measurements. Section 3 describes helicopter and fixed-wing AEM systems, and some of the transmitter-current waveforms as well as discussing the AEM footprint that is relevant for understanding the lateral resolution. Section 4 provides some examples of the results of AEM bathymetry studies from helicopter[3] time-domain and frequency-domain surveys in

[3] Refer to Vrbancich et al., (2005a,b) and Wolfgram & Vrbancich (2007) for examples of fixed-wing AEM bathymetry studies.

Australian waters, comparing the interpreted water depths with known values. Section 5 (Conclusion) summarises the important findings and briefly discusses future directions.

2. Development of the AEM bathymetry method – initial studies

Initial studies in the use of AEM for bathymetric mapping took place in the 1980s. Morrison and Becker (1982) were the first to consider the use of AEM systems for mapping water depths in a feasibility report commissioned by the US Office of Naval Research (ONR) to support rapid airborne bathymetric mapping in shallow coastal waters. Morrison and Becker (1982) investigated both time domain and frequency domain systems existing at the time and concluded that frequency-domain systems would be limited to a depth of approximately 20 m and that the INPUT[4] fixed-wing time-domain system, with a lower operating base frequency, was suitable for measurement depths of 50 to 60 m. This feasibility study was followed by single flight line field trials in Nova Scotia and New Brunswick (Canada) using the INPUT system over waters that were predominantly less than about 20 m in depth, resolving water depths to about 40 m with approximately 2 m accuracy as determined by comparison with soundings on coastal charts (Becker et al., 1984; Becker et al., 1986; Zollinger, 1985; Zollinger et al., 1987).

During this period of study, Won and Smits (1985, 1986b, 1987a), and Son (1985), also investigated the use of helicopter frequency-domain AEM for bathymetric applications. Based on field trials conducted in the Cape Cod Bay area (Massachusetts, USA) using a DIGHEM^{III} AEM system (Fraser, 1978), Won and Smits (1985, 1986b, 1987a) showed that excellent agreement in water depths was obtained with acoustic profiles down to depths of about 13 to 20 m corresponding to one and one and a half skin depths at 385 Hz (the lowest of the 2 frequencies used in the DIGHEM^{III} survey). The same dataset was also used to derive continuous profiles of seawater and seabed conductivity (Won & Smits, 1986a, 1987b). Bergeron and co-workers (1989) examined the same Cape Cod Bay dataset using a different method to invert the AEM data, to obtain good agreement with measured altimetry, seawater conductivity and known water depths. Bryan et al. (2003) used field data obtained over marsh and estuarine waters in Barataria Basin (Louisiana, USA) with a frequency domain AEM system using a primary waveform digitally constructed from six harmonic frequencies (Mozley et al., 1991a,b). The results of their inversions showed good agreement with measured water conductivities and depths, identifying horizontal water layers with the less saline, less conductive, fresher water layer overlying a more saline, more conductive water layer. The same instrumentation was used by Mozley et al. (1991a,b) in field trials at Kings Bay (Georgia, USA) to successfully map seawater depths and conductivity, and variations in seafloor sediment conductivity, and by Pelletier and Holladay (1994) to map bathymetry, sediment and water properties in a complex coastal environment located at Cape Lookout, North Carolina, USA.

2.1 Airborne electromagnetic bathymetry and sea ice thickness measurements

The EM response is sensitive to sub-metre variations in the altitude of the transmitter and receiver loops above the conductive seawater layer. This sensitivity has implications for measuring sea ice thickness. If the AEM data is also inverted for altitude, then the thickness

[4] INduced PUlse Transient; a trademark of Barringer Research Ltd.

of any sea ice covering the seawater can be determined by subtracting the elevation above the surface of the relatively resistive sea ice (measured with laser altimetry) from the altitude above the seawater (Becker & Morrison, 1983; Kovacs & Holladay, 1990; Reid et al., 2003a; Haas et al., 2009). This method relies on the conductivity contrast between sea ice and seawater. As sea ice ages, brine is expelled decreasing its conductivity. The seawater conductivity is typically 2.5 - 2.6 S/m, approximately two orders of magnitude larger than sea ice conductivity and at the relatively low frequencies used in frequency-domain AEM systems (typically 50 Hz to 100 kHz), the primary and induced secondary magnetic fields effectively "see through" the sea ice. The AEM system can therefore be used to measure sea ice thickness and also sub-ice bathymetry (Kovacs & Valleau, 1987; Pfaffling & Reid, 2009) in shallow waters. This same EM induction technique can also be applied using ship-borne sensors (Reid et al., 2003a; Reid et al., 2003b; Haas et al., 1997). Unlike the AEM bathymetry method which usually assumes a layered earth model for 1D inversion, determining the structure of three-dimensional sea ice keels (pressure ridges below the sea surface) with thicknesses that may reach 6 to 10 m may require the use of 2D and 3D EM modelling and inversion procedures (Reid et al., 2003a; Liu & Becker, 1990; Liu et al., 1991; Soininen et al., 1998) in order to minimise underestimating the thickness of the ice keels that results from smoothing of the EM response over the AEM system footprint (Kovacs et al., 1995; Liu & Becker, 1990) and interpretation using 1D inversion models.

3. AEM systems: Time-domain and frequency-domain

AEM systems operate in the frequency domain (frequency electromagnetic, FEM) or in the time domain (transient electromagnetic, TEM)[5]. Helicopter AEM systems typically operate in either the frequency domain or time domain, whilst fixed-wing AEM systems typically operate in the time domain. Fountain (1998) has reviewed the first 50 years of development of AEM systems from a historical perspective. A receiver loop is used to detect the secondary magnetic field induced in the ground. The recorded response is a voltage proportional to the time derivative of a component of the secondary field, typically the vertical component (dB_z/dt). In the following text, the term "bird" refers to the AEM system towed as a sling load beneath the helicopter in FEM and TEM systems, or to the receiver unit towed beneath a fixed-wing TEM system. All three types of AEM systems have been used for bathymetric investigations in Australian coastal waters.

3.1 Frequency domain helicopter AEM

The FEM system consists of several transmitter-receiver coil pairs held in a fixed rigid geometry. The transmitter-receiver coil pair separation is about 8 m or less, and the coils may be (i) placed horizontally (horizontal co-planar (HCP) configuration, vertical magnetic moment associated with the transmitter coil), (ii) or the coil pair may be rotated by 90°, so that the coils lie in the vertical plane (vertical co-planar (VCP) configuration, horizontal magnetic moment, orthogonal to the flight direction, associated with the transmitter coil), (iii) or the coil

[5] The AEM systems GEOTEM, DIGHEM, RESOLVE, HELITEM, and HeliGEOTEM (discussed in the following sections) are trademarks of Fugro Airborne Surveys Pty Ltd. QUESTEM was operated by World Geoscience Corporation since 1990 and became obsolete in 2000 after the amalgamation of World Geoscience Corporation and Geoterrex-DIGHEM Pty Ltd into the newly formed company Fugro Airborne Surveys Pty Ltd.

axes lie on the same horizontal line (vertical co-axial (VCX) configuration, horizontal magnetic moment in-line with flight direction, associated with the transmitter coil). Thus with respect to the flight direction, the HCP, VCP and VCX coil configurations have the transmitter dipole moment aligned vertically, transversely and longitudinally respectively with regards to the flight direction. The AEM data consists of the in-phase (*R*, real response) and quadrature (*Q*, imaginary response, relative to the primary field from the transmitter) signals detected by the receiver coils. The phase (φ) of the response is given by $\varphi = \arctan(R/Q)$.

The first AEM bathymetry survey in Australia took place in 1998 over Sydney Harbour (New South Wales, Australia) using a DIGHEM^V system (Figure 2a), originally developed by Fraser (1978) for resistivity mapping of metallic mineral deposits in the 1970s using earlier DIGHEM versions. The DIGHEM^V system is an analogue instrument[6] consisting of 5 coil pairs (3 HCP and 2 VCX, Table 1). The lowest frequency (*f*) was tuned to 328 Hz to

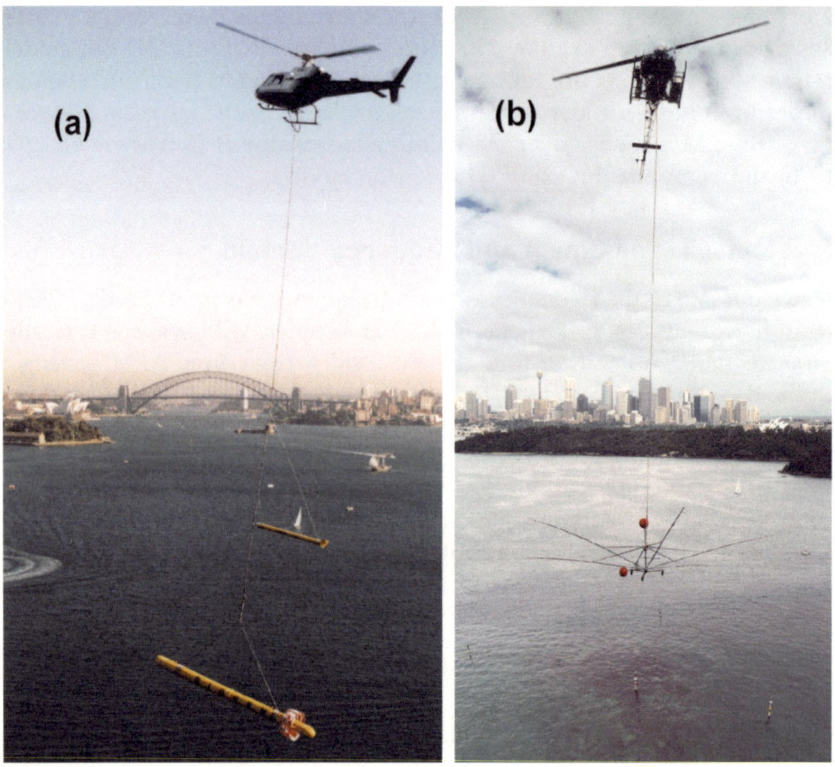

Fig. 2. (a): the frequency-domain helicopter DIGHEM^V AEM system (~ 8 m length) during survey over Sydney Harbour (Vrbancich, et al., 2000a,b). The smaller bird between the helicopter and the DIGHEM bird is a magnetometer bird; (b): The HoistEM time-domain helicopter AEM bird, located over the Sow and Pigs reef during a survey of Sydney Harbour (Vrbancich & Fullagar, 2004, 2007b). The transmitter loop (~ 22 m diameter) is attached to the extremities of the poles and the multi-turn receiver loop is located on the same plane, at the centre of the system.

[6] The current system is the digital RESOLVE system, used in Australia for example to study salinisation, e.g. Hatch et al., (2010).

maximise the depth of penetration through seawater, i.e., to increase the skin depth (δ) where δ (m) = $500/(\sigma.f)^{1/2} = 250/(f)^{1/2}$ for a typical seawater conductivity (σ) of 4 S/m. Following this survey, Shoalwater Bay (Queensland, Australia; Vrbancich, 2004) and Sydney Harbour have been surveyed using an analogue DIGHEM_Res(istivity) instrument with 5 HCP coil pairs operating within the range of 387 Hz to 103 kHz (Table 1).

	f1 (Hz)	f2 (Hz)	f3 (Hz)	f4 (Hz)	f5 (Hz)
DIGHEM(V)	328 HCP	889 VCX	5658 VCX	7337 HCP	55300 HCP
Skin Depth: δ (2δ)	13.8 (27.6)	8.4 (16.8)	3.3 (6.6)	2.9 (5.8)	1.1 (2.2)
Resistivity Bird	387 HCP	1537 HCP	6259 HCP	25800 HCP	102700 HCP
Skin Depth: δ (2δ)	12.7 (25.4)	6.4 (12.8)	3.2 (6.4)	1.6 (3.2)	0.8 (1.6)

Table 1. Frequencies and associated skin depths (m) for the DIGHEM^V and DIGHEM-Resistivity AEM birds. Skin depth δ (m) assumes a typical seawater conductivity of 4 S/m. Depth of investigation is equivalent to approximately 2δ (m).

One advantage of FEM helicopter AEM systems (compared to fixed-wing TEM systems) is that the transmitter and receiver coils are contained in a rigid structure so that the coils are held in a fixed position relative to each other. However pendulum motion of the towed bird generates a geometric and inductive effect in the measured EM response and contributes to altimeter error (Davis et al., 2006). If the bird swing period can be determined from survey data, a filter can be designed to remove bird swing effects, to first order, caused by pendulum motion (Davis et al., 2009). Predictions of bird swing from GPS receivers mounted on the bird and the helicopter can be used to predict the bird maneuver (Davis et al., 2009; Kratzer & Vrbancich, 2007). Calibration errors in FEM helicopter data can be identified by transforming the data to several different response-parameter domains and used to minimise the effect of altimeter and bird maneuver errors (Ley-Cooper et al., 2006). Importantly, this procedure can be applied to historic data, i.e., previous/dated FEM helicopter AEM data can be re-analysed with this procedure.

3.2 Fixed-wing time-domain AEM

A TEM system does not use a continuous sinusoidal current to power the transmitter loop (as in FEM systems). Instead, typically, the transmitter current is increased to a maximum value (during the "on-time"), and then reduced to "zero" current and measurements are made during the "off-time" when the transmitter is not powered[7]. Two examples of TEM waveforms are shown schematically in Figure 3. The secondary magnetic fields related to the decay of the induced currents in the ground are detected whilst there is no primary field. The process is repeated using the same current pulse but with the opposite polarity and the same off-time interval (Figure 3), and this response is subtracted from the first response to

[7] On-time measurements are also possible with some TEM systems but are not discussed here. The TEM fixed-wing and helicopter TEM systems used for AEM bathymetry studies in Australia use data recorded only during the off-time.

improve the signal to noise ratio. The period of the waveform (consisting of the the two on-time periods (Δt_1) of opposite polarity and their corresponding off-time periods, Δt_2) represents the base frequency($1/\Delta t_3$), Figure 3; 25 Hz is used in Australia and 30 Hz is used in America to minimise 50 Hz and 60 Hz power line transmission interference respectively. The results of the subtraction between the measurements made during the first and second halves of the waveform are stacked (averaged) over many cycles to reduce noise.

The shape of the recorded waveform (transient decay, Figure 3c) during the off-time is equivalent to the response at a number of frequencies (ranging high to low) for a harmonically varying source. Thus different sections of the decay curve contain different proportions of high and low frequency components. Morrison et al. (1969) computed the vertical component of the transient field from an airborne horizontal loop above a layered ground and observed that at early times (shortly after the transmitter current has dropped to zero), the response is due to both high and low frequency components whilst at later times approaching the end of the off-time interval, only the low frequency response

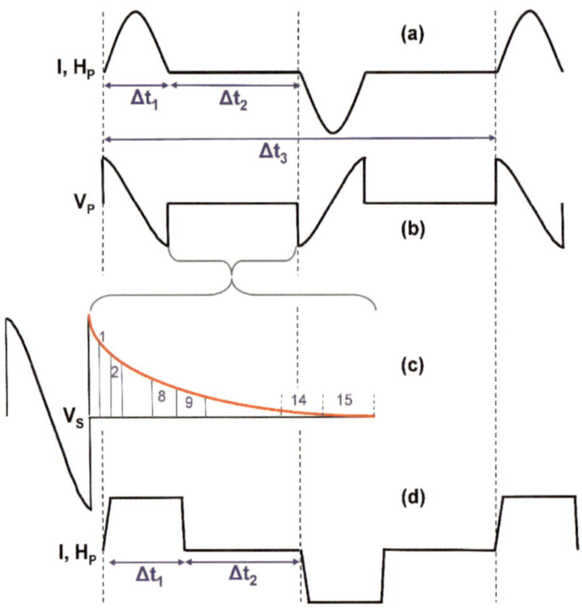

Fig. 3. Current and voltage waveforms. (a): Half-sine bipolar wave pulse – transmitter current (I) and primary magnetic field (H_P) waveform (e.g. GEOTEM, QUESTEM). On-time (Δt_1) typically 4 ms, off-time (Δt_2), typically 16 ms. For a base frequency of 25 Hz ($\Delta t_3 = 40$ ms), $\Delta t_1 + \Delta t_2 = 20$ ms (half period). (b): Voltage (V_P) from primary field at the receiver loop, or at high altitude where there is no response from the ground, i.e., no voltage (V_S) from secondary magnetic fields. (c): Response from ground (V_S) during off-time: transient decay curve (EM response) shown in red, sampled typically 120 to 250 times and binned into typically 15 to 30 windows (approximately logarithmically spaced) with narrow windows at early times to capture rapid decay and wider windows at late times to capture slower decays. (d) Quasi-trapezoidal current waveform (e.g. HoistEM, RepTEM, SeaTEM), 25 Hz base frequency, on-time (Δt_1) typically 5 ms, off-time (Δt_2), typically 15 ms, typically 21 to 28 windows. Not to scale.

remains. Morrison et al (1969) also noted that given that the skin depth is inversely proportional to the square root of the frequency, then the early part of the transient decay is governed by rapid decay of high frequency energy which has only penetrated down to relatively shallow depths whilst at later times, the response is dominated by lower frequency energy which has penetrated to greater depths. Resistive media are associated with a rapid decay, conductive media are associated with a longer decay. For bathymetric applications, early time (and higher frequency) AEM data are required for shallow seawater depths and late time (and low frequency) AEM data are required for deeper seawater. In principle, the lower the base frequency for TEM systems and the lower the frequency for FEM systems, then the greater the depth of investigation in seawater.

3.2.1 Variable transmitter-receiver geometry

Fixed-wing TEM systems have the transmitter coil spanning the wingtips and front and rear extremities of the aircraft, as shown schematically in Figure 1, with the receiver coil contained in a "bird" that is released from the rear of the aircraft. The transmitter and receiver operate at different heights above ground level. The aircraft survey altitude is typically about 120 m and the bird has an assumed fixed horizontal offset within the range of 90 to 120 m and an assumed fixed vertical offset within the range of 40 to 60 m, relative to the centre of the transmitter. Unlike FEM helicopter systems, the relative geometry between the receiver and transmitter is variable leading to interpretation errors arising from unrecorded variations of bird attitude, offset and altitude, thereby limiting the potential of the AEM bathymetry method. Vrbancich & Smith (2005) estimated bird position errors of several metres at survey altitude using GEOTEM data, based on the prediction of the vector components of the primary field measured by the receiver at high altitude, i.e. assuming a free-space approximation (Smith, 2001a) and at survey altitude over conductive seawater by estimating the distortion of the primary field caused by an in-phase contribution to the primary field from the seawater response (Smith, 2001b). These approximate methods for determining bird position only have an accuracy of a few metres and may therefore be limited. Another method involving the determination of bird position and receiver coil attitude as parameters obtained in a least-squares sense during layered-earth inversion of multi—component datasets (Sattel et al., 2004) has been applied to GEOTEM data to significantly improve the bathymetric accuracy of an AEM bathymetry survey in Torres Strait, located between Australia and Papua New Guinea (Wolfgram & Vrbancich, 2007).

3.3 Helicopter time domain AEM

Helicopter TEM systems are similar to the fixed-wing systems except that both the transmitter and receiver loops are located on a framework suspended as a sling load beneath the helicopter, as shown in Figure 2b. The waveforms are similar to those of fixed-wing TEM systems, i.e., an on-time period followed by an off-time period (Figure 3). Sattel (2009) has reviewed current helicopter TEM systems. The VTEM (Geotech Ltd.; Witherly, 2004) and HeliGEOTEM/HELITEM (Smith et al., 2009) systems for example are typically used for mineral exploration and the SkyTEM system (Sorensen & Auken, 2004) was developed for hydrogeophysical and environmental applications. The HoistEM (Boyd, 2004) and RepTEM systems have been used for several AEM bathymetry surveys in Australian coastal waters (e.g. Sections 4.2-4.4), together with the SeaTEM system (e.g. Section 4.5) which was developed alongside the RepTEM system[8] specifically for bathymetric applications. An example of the SeaTEM waveform is shown schematically in Figure 3d. The HoistEM, RepTEM and SeaTEM systems are central loop systems with the receiver loop and transmitter loops co-axially aligned and lying within the same plane (i.e., nominally no vertical separation between the loops).

3.4 AEM footprint

The AEM footprint is a measure of the lateral resolution of an AEM system and was originally studied for frequency-domain AEM in the context of sea ice measurements by Liu & Becker (1990). The AEM footprint, in the inductive limit, was defined as the square area centred beneath the transmitter that contained the induced currents responsible for 90% of the observed secondary magnetic field detected at the receiver (Liu & Becker, 1990). The inductive limit refers to the case of a perfect conductor and/or infinite transmitter frequency, and as such, the induced currents are entirely in-phase with the primary field (i.e. no quadrature component) and the Liu-Becker footprint therefore corresponds to the minimum in-phase footprint of a frequency-domain AEM system. In the case of finite transmitter frequency and earth conductivity, the currents induced in the earth will have a larger spatial extent than that at the inductive limit (Beamish, 2003) and will contain both in-phase and quadrature components. Reid & Vrbancich (2004) have compared the inductive-limit footprints of various AEM configurations including time-domain systems in order to analyse the suitability of AEM systems for Antarctic sea-ice thickness measurements. Reid et al. (2006) extended the Liu-Becker footprint calculations (Liu & Becker, 1990; Reid & Vrbancich, 2004) to the case of finite frequency and earth conductivity for HCP and VCX configurations over an infinite horizontal thin sheet and for a homogeneous half-space. This study found that AEM footprint sizes may be several times the Liu-Becker inductive-limit value, with the quadrature footprint approximately half to two-thirds that of the in-phase footprint.

The Liu-Becker footprint is determined by both the transmitter-receiver geometry and the altitude (h). For a dipole-dipole frequency-domain AEM configuration with a transmitter-receiver separation of 6.3 m, the Liu-Becker footprint is 3.73h and 1.35h for HCP and VCX coil geometries. For a central loop system (e.g. HoistEM), the footprint is 3.68h, see Table 2 for a comparison of footprint sizes between several AEM systems. Beamish (2003) also computed the AEM footprint, using a different definition based only on the current system induced in the earth by the transmitter neglecting the contributions the secondary magnetic fields induced by currents in the ground make at the receiver. Beamish determined footprint sizes of between 0.99h and 1.43h for a horizontal magnetic dipole source (transmitter), and between 1.3h and 2.1h for a vertical magnetic dipole source, depending on the transmitter frequency and half-space conductivity. Section 4.4 presents an example of how the EM footprint affects the lateral and vertical resolution using 1D inversion.

[8] The HoistEM system was developed by the Normandy Group and the RepTEM and SeaTEM systems were developed by Geosolutions Pty.Ltd. The RepTEM system recently replaced the HoistEM system.

System	Tx height	Rx height	Tx-Rx separation	Geometry	Footprint/Tx-height
DIGHEM	30 m	30 m	8.0 m	HCP	3.72
				VCX	1.34
				VCP	1.44
SkyTEM	30 m	30 m	0.0	Tx(z)-Rx(z)	3.68
GEOTEM	120 m	70 m	130.0 m	Tx(z)-Rx(z)	4.51
				Tx(z)-Rx(x)	2.97

Table 2. Liu-Becker inductive-limit footprint sizes (footprint to transmitter height ratios) for several common AEM systems. Tx: transmitter, Rx: receiver. For TEM systems, the geometry is specified by the directions of the Tx and Rx dipole axes: Tx(z)-Rx(z) denotes a vertical-axis Tx and vertical-axis Rx, Tx(z)-Rx(x) denotes a vertical axis Tx and a horizontal-axis (inline) Rx. For the central loop helicopter TEM configuration (e.g. SkyTEM, HoisTEM), the Tx loop was a finite horizontal loop (10 m x 10 m) and the Rx was a vertical axis point dipole. Adapted from Table 1, Reid & Vrbancich (2004).

4. AEM bathymetry studies in Australian coastal waters

In 1998, the Australian Hydrographic Service (formerly known as the Royal Australian Navy Hydrographic Service) and the Defence Science & Technology Organisation (DSTO) began investigating the use of AEM as a rapidly-deployable bathymetric mapping technique in Australian waters,[9] complementing the use of lidar. Commercial systems were initially used from 1998 to 2005 for field trials to appraise the accuracy of the AEM bathymetry method based on (i) time-domain fixed-wing AEM using the GEOTEM and QUESTEM systems (Vrbancich et al., 2005a,b; Wolfgram & Vrbancich, 2007), (ii) time-domain helicopter AEM using the HoistEM system (Vrbancich & Fullagar, 2004, 2007b) and (iii) frequency-domain helicopter AEM using the DIGHEM[V] system (Vrbancich et al., 2000a,b) and DIGHEM_Resistivity system (Vrbancich, 2004). Between 2006 and 2010, field trials were conducted with a prototype SeaTEM system (Vrbancich, 2009), a floating system equivalent to the commercial RepTEM system (Vrbancich et al., 2010) and the SeaTEM system (Vrbancich, 2011; Vrbancich, 2010). The SeaTEM system is essentially a slightly smaller version of RepTEM, developed for DSTO by Geosolutions Pty Ltd. as a research instrument. The results of the floating AEM system were significant, providing an upper limit to the accuracy expected from AEM systems similar to the RepTEM central loop system.

4.1 Sydney Harbour – helicopter frequency-domain AEM (DIGHEM)

A DIGHEM[V] survey undertaken in 1998 (Vrbancich et al., 2000a) was flown over the width of the entrance to Sydney Harbour including the Sow and Pigs reef located in the centre of the channel. The nominal bird height was 30 m. Two interpretation methods were used to

[9] The scope for testing the potential use of AEM for bathymetric mapping is significant, especially within waters affected by turbidity and in the surf zone. Within the Australian Charting Area, the area of waters less than 70 m in depth that are considered to be *well* surveyed by any method, is estimated to be about 599,000 square kilometres. The areas of waters less than 50 m and between 50 to 70 m in depth that are considered to be *poorly* surveyed are estimated to be about 531,000 and 867,000 square kilometres respectively. It is also estimated that approximately 30% of the waters within the 70 m depth contour are affected by turbidity. (Australian Hydrographic Service, personal communication, 2010.)

estimate the water depths: direct layered-earth (1D) inversion and the conductivity-depth transform (CDT). The former method uses program *AEMIE* (Fullagar Geophysics Pty Ltd) based on a modified version of a 1D inversion program for horizontal loop EM (Fullagar & Oldenburg, 1984) and the latter method is a rapid approximation method not relying on inversion, to achieve fast processing and interpretation (Macnae et al., 1991; Wolfgram & Karlik, 1995; Fullagar, 1989) using program *EMFlow* (Macnae et al., 1998). The concept of the conductivity-depth transform is crudely described as follows. For each time in the transient decay (Figure 3c), the conductivity of a homogeneous half-space is computed such that the predicted magnetic field amplitude matches the observed field at that given time. Thus providing the field amplitude at a series of times gives the apparent conductivity[10] (σ_a) at each time *t*. This process associates a set of apparent conductivities with the decay times ($\sigma_a \Leftrightarrow t$), and the depths (*d*) of the "smoke ring" diffusing through the homogeneous ground as a function of time is associated with a set of times ($d \Leftrightarrow t$), thus the apparent conductivity is associated via time with depth (conductivity-depth transform, $\sigma_a \Leftrightarrow d$).

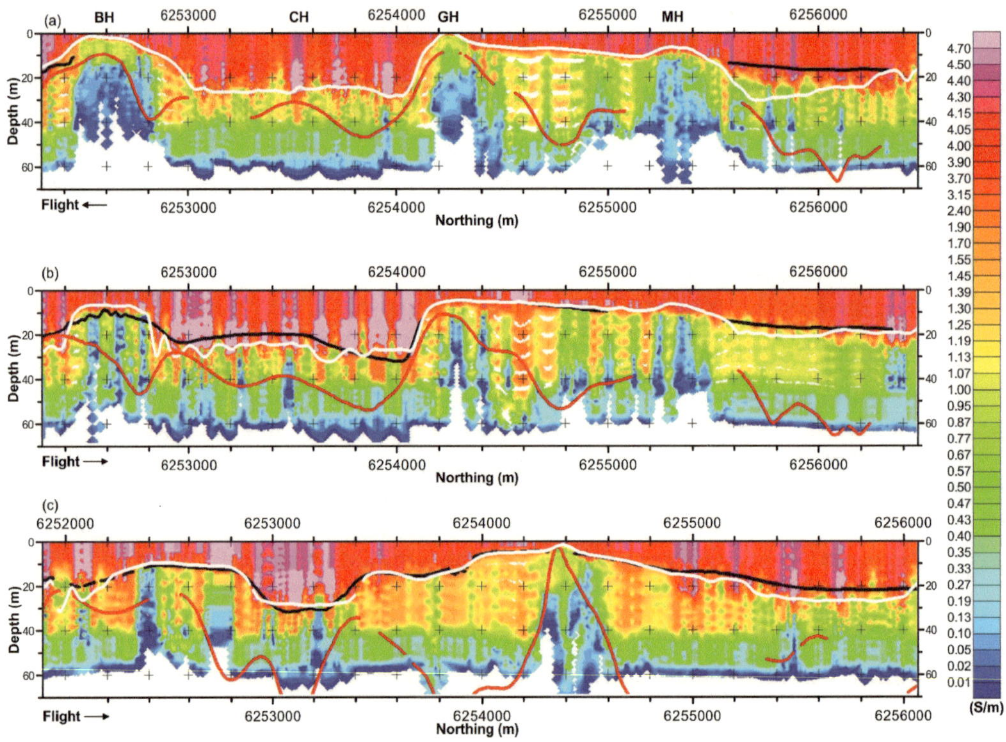

Fig. 4. *AEMI* (layered-earth inversion) conductivity sections (S/m) with profiles of depth to bedrock from seismic surveys (red), echo sounding water depths (black) and Geoterrex-DIGHEM proprietary inversion software (white). (a): this line skirts the coastline and lies closest to the headlands, marked BH, CH, GH and MH; (b) and (c) progress further away from the coastline; line spacing is nominally 50 m. Source: Vrbancich et.al., 2000b.

[10] The apparent conductivity is the conductivity of a homogeneous half-space which will give rise to the same EM response as that that would be measured over a real earth.

The results of a layered earth inversion for three adjacent flight lines that flank four sandstone headlands are shown in Figure 4. For this inversion, four layers were assumed in the starting model overlying a resistive basement. If conductivities greater than about 2.5 S/m are attributed to seawater, the inverted conductivity-depth sections are in good agreement with known water depths, even in the deeper regions of ~ 30 m. The equivalent conductivity-depth sections obtained from the conductivity-depth transform using program *EMFlow* is shown in Figure 5. (Both Figures 4 and 5 use the same colour scale bar.) The interpreted seawater layer has a higher average conductivity than the layered earth inversion, typically about 6 S/m (measured conductivity was 4.5 - 4.6 S/m). Generally however, there is very good agreement with measured water depth soundings, especially down to about 20 m. This comparison shows that the conductivity-depth transform method can provide useful bathymetric data quickly, but the interpreted water depths for this dataset are not quite as accurate as those obtained from layered-earth inversion.

The features in Figures 4, 5 that are less conductive than seawater and more conductive than the resistive bottom layer (i.e., coloured orange-yellow-green) are indicative of unconsolidated marine sediments. The very shallow marine sediments indicating bedrock at or near the seafloor is clearly identified where the survey line is closest to the headlands

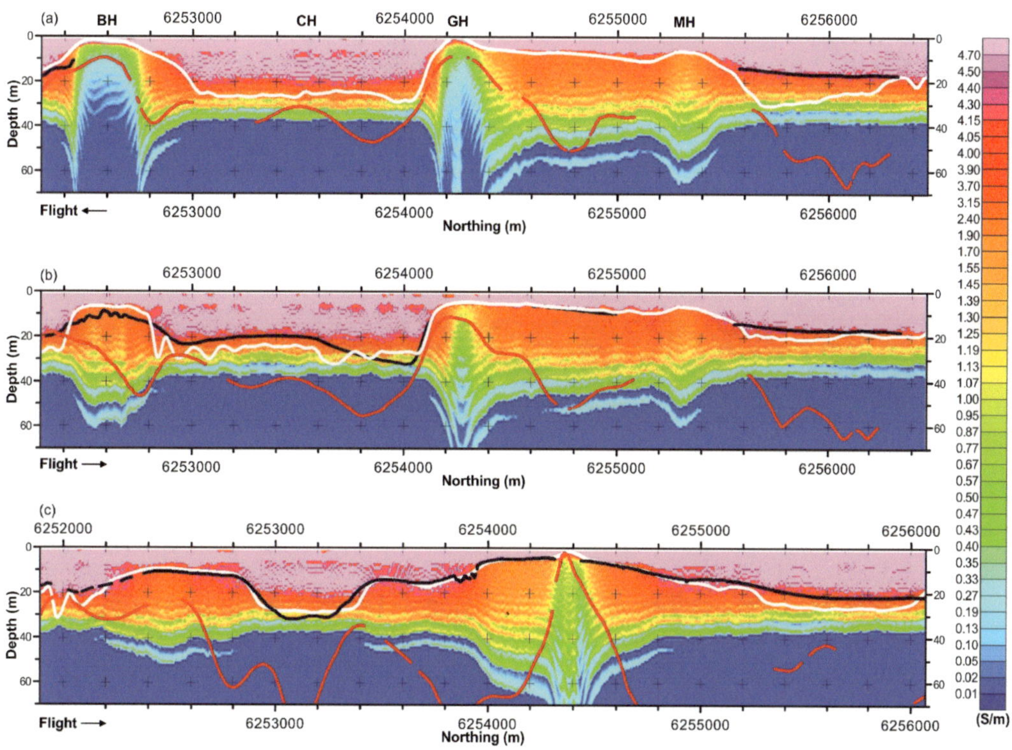

Fig. 5. *EMFlow* (conductivity-depth transform) conductivity sections (S/m) with profiles of depth to bedrock from seismic surveys (red), echo sounding water depths (black) and Geoterrex-DIGHEM proprietary inversion software (white). (a): this line skirts the coastline and lies closest to the headlands, marked BH, CH, GH and MH; (b) and (c) progress further away from the coastline; line spacing is nominally 50 m. Source: Vrbancich et.al., 2000b.

marked BH, GH and MH in Figures 4a and 5a. The seismic profiles (shown in red) were obtained from an earlier study (Emerson & Phipps, 1969). Another marine seismic study (Harris et al., 2001) confirms that the bedrock reaches the seafloor adjacent to the headland MH at a depth predicted by the AEM data as shown in Figures 4a, 5a, and that the seismic line in Figures 4b, 5b adjacent to the headland BH is incorrect and that the actual bedrock contour in this region is in agreement with that predicted by the AEM data (Figure 5b).

4.2 Sydney Harbour – helicopter time-domain AEM (HoistEM)

Sydney Harbour was resurveyed again in 2002 using the HoistEM time-domain helicopter AEM system (Figure 2b). The AEM data were improperly calibrated (Vrbancich & Fullagar, 2004) and was corrected using a procedure that reconciles the measured data with available ground truth at selected points within the survey (Vrbancich & Fullagar, 2007b). An example of a conductivity section resulting from a line that was flown over the Sow and Pigs reef is shown in Figure 6. A two-layer-over-basement model was used for the inversion. The fitted upper layer conductivity agrees well with the measured seawater conductivity and the seawater-sediment interface accurately follows the known water depth profile (yellow), mostly to within 1 m or better. The maximum depth of investigation was found to be ~ 55 - 65 m, just offshore of the harbour entrance. The time-domain equivalent of the skin depth is the diffusion length (δ_{TD}) given by $\delta_{TD} = \sqrt{2t/\sigma\mu_0}$ where σ is the conductivity (S/m) and μ_0 is the magnetic permeability of free space. Assuming a seawater conductivity of 4 to 5 S/m, δ_{TD} corresponds to a depth of 76 to 68 m respectively at the late time (t) of 14.4 ms and this represents an estimate of the maximum depth of investigation for this AEM system. Vrbancich & Fullagar (2007a) also showed that this dataset, and hence the AEM technique, has the potential to identify the coarse features of the underlying bedrock topography, as partly revealed in Figure 6.

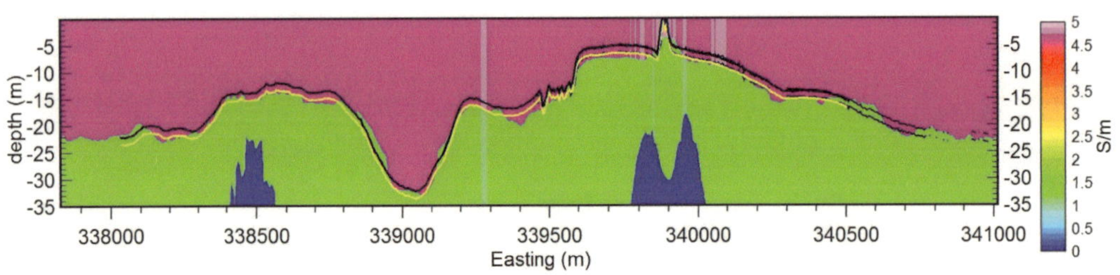

Fig. 6. Conductivity-depth sections from 1D inversion using corrected data to account for imperfect AEM instrument calibration. Accurate bathymetry profiles (obtained from combined multi-beam sonar and lidar data) are shown: (i) tide corrected (black) referred to a given datum; (ii) actual water depth at time of survey (yellow), which includes a tide height of 1.4 m. The Sow and Pigs reef is shown as a pinnacle at ~ 339900 m (E). The colour scale has units of S/m. Measured seawater conductivity is ~ 4.7 S/m. The green section represents sediment and the blue sections represent underlying resistive basement (bedrock). Source: Vrbancich & Fullagar, 2007b.

Figure 7 compares 3D images of the seafloor topography (defined by the water depth); the lower image is derived from inversion of corrected AEM data and was obtained by interpolating the water depth profiles defined by the seawater-sediment interface depth, as shown Figure 6, to a gridded surface (i.e. by gridding the depth profiles). The upper image was obtained by gridding the known water depths. For the purpose of this comparison, individual known bathymetry grids from several high-spatial density sonar surveys and airborne laser depth soundings were combined into a single grid and sampled at the HoistEM measurement locations (fiducial point coordinates) to produce bathymetric data at the relatively low spatial density of the AEM data, e.g., the black and yellow curves in Figure 6. Gridding the series of yellow curves across all the survey lines yields the upper image in Figure 7.

4.3 Crookhaven Bight – helicopter time-domain AEM (SeaTEM)

Figure 8 shows a representative example of a profile (line L1185, red) of interpreted water depths obtained from layered-earth inversion of SeaTEM data, from a survey flown over Crookhaven Bight (~ 35.0° S, 150.8° E) in November 2009, adjacent to Jervis Bay and located approximately 150 km south of Sydney. The western section of the profile joins Currarong Beach and this section and the coastal area adjacent to the beach follows the seafloor gradient accurately (compare with black and grey profiles obtained from sonar data) to a water depth of about 25 to 30 m. The peaks at ~ 301400 and 302300 m (E) arise from reef structure, identifying a cross-section of the "zig-zag" nature of palaeochannels incised into the bedrock, as shown by the troughs located at 301800 and 301300 m (E) (Vrbancich et al., 2011).

Fig. 7. Three-dimensional images of the seafloor topography based on pre-existing sonar-lidar data (top) and on inversion of corrected HoistEM data (bottom). Inverted seawater depths were not smoothed prior to gridding. The images are vertically aligned and are depicted with the same vertical exaggeration and colour scale. The surface is coloured according to water depth: red (0-4 m); orange (< 7m); yellow (<10 m); pale green (< 16 m); aqua (< 23 m) and dark blue (< 32 m). The "white" arrowhead (top) marks the location of the Sow and Pigs reef. Source: Vrbancich & Fullagar, 2007b.

The AEM survey was flown using 30 lines with 100 m line spacing. A colour-shaded surface grid obtained by gridding all the 30 profiles is shown in Figure 9 and can be compared directly with an equivalent grid of the single-beam sonar data, shown in Figure 10. All of the significant features and most minor features are accurately identified in the AEM

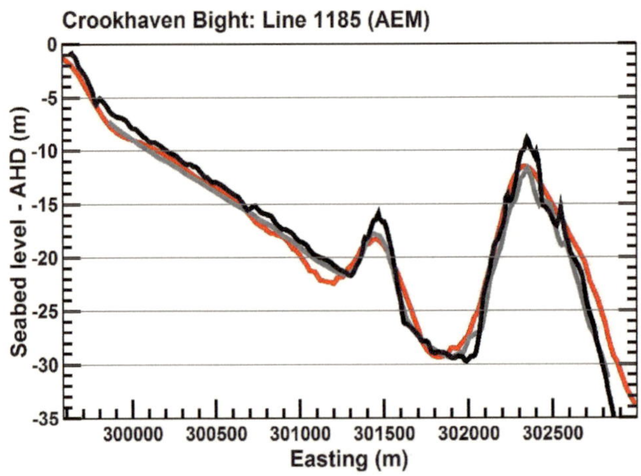

Fig. 8. Seabed level relative to Australian Height Datum (AHD) for survey line L1185; red: AEM bathymetry; black: Australian Hydrographic Service single-beam sonar data; grey: single-beam sonar data recorded during marine seismic survey (Vrbancich et al., 2011a). The location of this profile is shown in Figures 9 and 10.

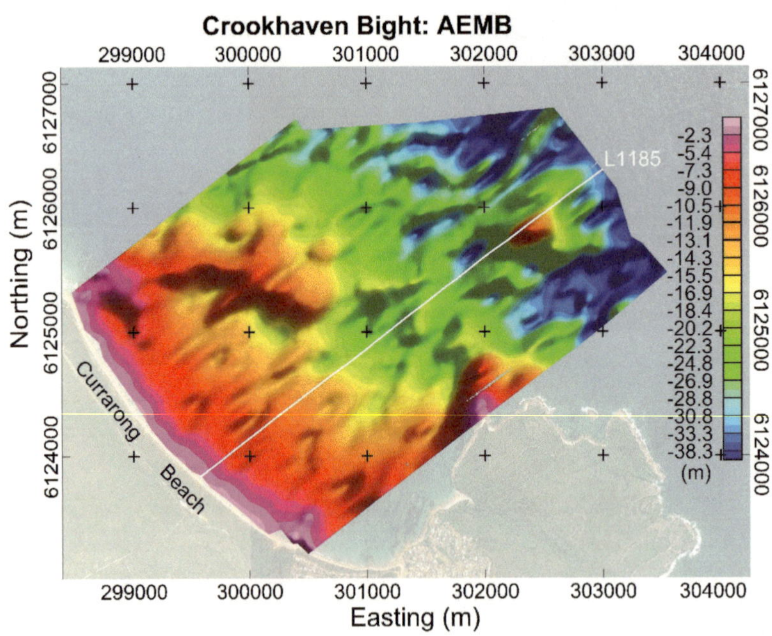

Fig. 9. Bathymetry of Crookhaven Bight survey area derived from AEM (SeaTEM) data. The bathymetry profile for line L1185 is shown in Figure 8 (red). The colour scale refers to the water level (m) relative to Australian Height Datum.

bathymetry map which took approximately 2 hours to survey and about 1 hour to process the data and run the inversions to produce a georeferenced map. The SeaTEM AEM system is experimental. In some regions, as shown in Figure 9, the interface depths display an oscillatory profile in some areas which degrades the accuracy of the derived water depths. Intensive tests conducted after this survey have shown that these anomalies may be caused by a high frequency noise source which cannot be removed by the usual data processing procedures. The source of this noise is currently being investigated.

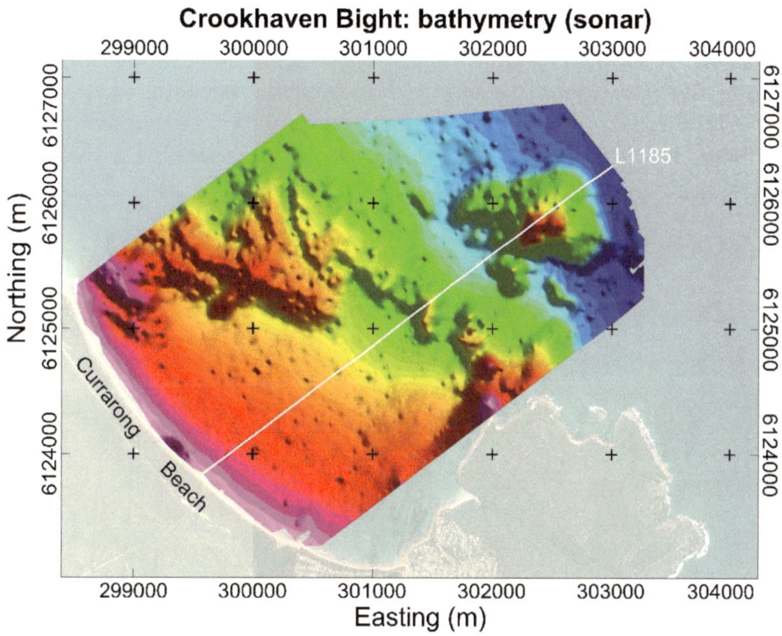

Fig. 10. Bathymetry of Crookhaven Bight survey area derived from sonar (Australian Hydrographic Service) data. The bathymetry profile for line L1185 is shown in Figure 8 (black). The colour scale is the same as shown in Figure 9.

4.4 Yatala Shoals – helicopter time-domain AEM (HoisTEM) – AEM footprint features

Yatala Shoals (~ 35.7° S, 138.2° E) is located in Backstairs Passage, South Australia, between the Fleurieu Peninsula and Kangaroo Island. This area is interesting because it contains a series of ridges, which fan out, increasing in separation and decreasing in height above the seafloor. This area was surveyed in 2003 using the HoistEM system, to examine the effect of the EM footprint on the lateral resolution, for resolving peak separation and height.

Figure 11 shows the results of layered-earth inversion for two profiles, separated by about 300 m. The main feature is a ridge rising approximately 18 m from the seafloor at ~ 244700 m (E) (Figure 11a) which gradually deepens and splits into two ridges (244390 and 244670 m (E), Figure 11b). A series of narrow ridges and troughs flank the eastern side of this main ridge. The impact of the AEM footprint is evident as a smoothing and underestimation of peak heights. Referring to the AEM bathymetry profiles (red), the narrow trough (~ 100 m gap) to the west of the main ridge in Figure 11a between 244400 and 244700 m (E) is unresolved, yet as this ridge splits into two ridges with a trough wider than 300 m (Figure

11b), the height and width of the western ridge (~ 244390 m (E)) is well resolved, and the secondary peak at 244670 m (E) is clearly resolved, yet its width is over determined and the height of the ridge is noticeably underestimated by about 3 m. The ridges that flank the eastern side of the main ridge are relatively narrow (~ 100 to 150 m) and are smaller than the central loop EM footprint, thus whilst the peaks are identified, their elevations above the seafloor are all underestimated by several metres. Another factor contributing to the apparent smoothing of the peak structure is the effect of assuming a 1D (layered-earth) model for inversion. 2D/3D modelling and inversion would be more appropriate in this case, however the routine use of *generalised* 3D modelling and inversion methods has not been realised, furthermore, the application of existing methods to an entire AEM survey dataset is impractical because of the required computer processing time. Cox et al. (2010) have recently introduced a robust 3D inversion scheme based on a moving AEM footprint that would enable the effects of the 1D approximation (i.e. assuming a layered-earth) to be fully investigated.

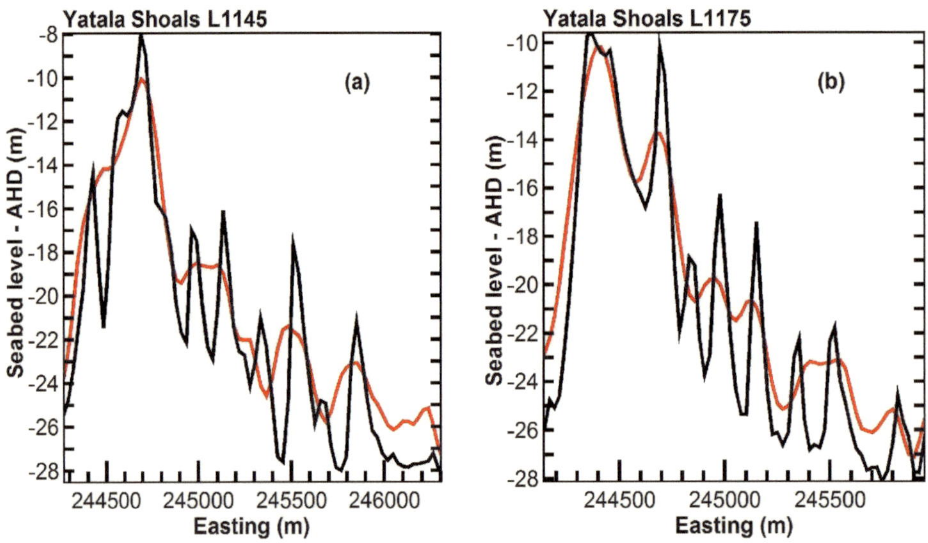

Fig. 11. Seabed levels, Yatala Shoals, relative to Australian Height Datum for HoistEM lines (a) L1145, (b) L1175, separated by approximately 300 m; red: AEM bathymetry; black: lidar.

4.5 Palm Beach (Broken Bay) – helicopter time-domain AEM (SeaTEM)

Figures 12a,b show a comparison of known bathymetry (sonar) and bathymetry derived from AEM data, obtained from a SeaTEM survey undertaken in 2009 over waters adjacent to Palm Beach, Broken Bay (~ 33.59° S, 151.33° E) located ~ 40 km north of Sydney Harbour. The depths were obtained using a two-layer-over-basement model using measured seawater conductivity, altimetry and a sediment conductivity of 1.25 S/m (consistent with sampled sediments in the region, Vrbancich et al., 2011b) as fixed parameters in the inversion. The AEM bathymetry map shows very good agreement with the known bathymetry, highlighting the gradual deepening of waters offshore Palm Beach, and the shoal region adjacent to Barrenjoey Head. A palaeochannel running approximately east-west was mapped, based on AEM data, about 60 m below marine sediments in this area, estimated to cut across Palm Beach at ~ 628225 m (N). The extent and depth of the palaeochannel in this area was found to be in very good agreement with a depth-to-bedrock map derived from marine seismic data (Vrbancich et al., 2011b).

5. Discussion and conclusion

The use of AEM methods, traditionally applied to mineral exploration, can be applied to the measurement of water depths and seawater and seabed conductivity in shallow coastal waters that may be turbid and lie within the surf zone, thereby limiting the application of lidar techniques in these waters. The pioneering work in this area was carried out in Canada and the USA, with most of this work published in the 1980s and early 1990s[11]. This paper presents some of the findings of AEM bathymetry studies in Australian coastal waters carried out between 1998 and 2010 using helicopter frequency- and time-domain systems. The AEM footprint (and hence the lateral resolution) is significantly greater than the lidar footprint, and this limits the AEM method for bathymetric mapping where International Hydrographic Organisation (IHO) standards are required. However sub-metre vertical water depth resolution can be obtained using AEM methods in shallow waters (within a 10 to 20 m depth range). Presently, the AEM bathymetry method is a reconnaissance technique that can be rapidly deployed in remote areas via helicopter, for fast estimation of coastal water depths, including waters in the surf zone and turbid waters, to depths of about 30 to 40 m, and for identifying areas of exposed reef and for estimating the coarse features of the underlying bedrock topography.

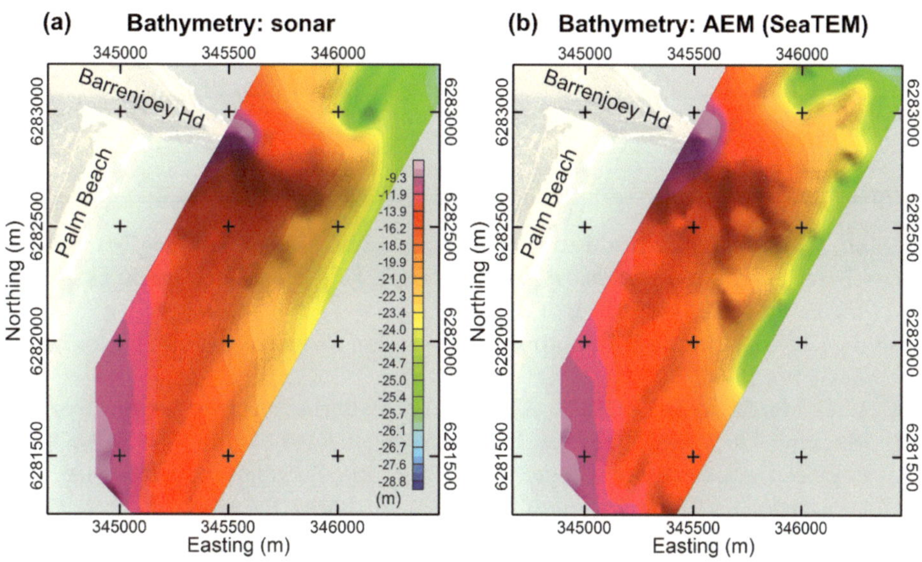

Fig. 12. Bathymetry maps of an area adjacent to Palm Beach, Broken Bay: (a), sonar; (b) AEM. Both maps use the same colour scale to show the seabed level relative to the Australian Height Datum.

[11] Many of the technical development issues, applications, recommended areas of research and strategies for further development, as discussed in a working group report in 1987 (Bergeron et al, 1987), are still relevant.

One of the most important issues is instrument calibration, affecting both environmental and exploration applications of AEM. Data obtained from (i) measurements of seawater depth and conductivity, (ii) sediment conductivity from bore hole samples and (iii) sediment depths estimated from marine seismic data, all acquired from a suitable trial site, can be combined to provide ground-truth data to establish a suitable geo-electrical model, for checking instrument calibration errors by comparing derived depths and conductivities with "known" depths and conductivities. Improved instrument calibration will lead to more robust estimates of seawater depth, seabed properties, sediment thickness and bedrock topography as well as improved depth accuracy and deeper depths of investigation.

Apart from improving AEM instrument calibration, reducing noise sources and including sensors to accurately track bird motion, future work will also involve software enhancements in program *AMITY* and associated software (Fullagar Geophysics Pty Ltd) to enable geologically-constrained 1D time-domain inversion, and 1D extremal inversion which enables models to be constructed with maximal and minimal characteristics (e.g. depth and conductivity bounds) which fit the observed data to within acceptable levels. Where warranted, future studies will also include the use of 3D EM modelling and inversion methods for interpretation of AEM data.

6. Acknowledgment

I thank Graham Boyd (Geosolutions Pty Ltd), Keith Mathews (Kayar Pty Ltd), Richard Smith (Technical Images Pty Ltd) and Peter Fullagar (Fullagar Geophysics) for their contributions to the AEM bathymetry (AEMB) studies in Australian waters. I thank the Australian Hydrographic Service for supporting AEMB and the Defence Imagery & Geospatial Organisation (DIGO) for release of aerial imagery under the NextView licence, as used in Figures 9, 10 & 12.

7. References

Archie, G.E. (1942). The Electrical Resistivity Log as an Aid in Determining Some Reservoir Characteristics. *Transactions American Institute for Mining, Metallurgical and Peroleum Engineers*, Vol. 42, pp. 54-62.

Beamish, D. (2003). Airborne EM Footprints. *Geophysical Prospecting*, Vol. 51, No. 1 (January 2003), pp. 59-60.

Becker, A. & Morrison, H.F. (1983). Analysis of Airborne Electromagnetic Systems for Mapping Thickness of Sea Ice. Prepared for NORDA, contract N62306-83-M-1755. Engineering Geoscience, University of California, Berkeley, November 1983.

Becker, A.; Morrison, H.F. & Zollinger, R. (1984). Airborne Electromagnetic Bathymetry. *SEG Expanded Abstracts*, Vol. 3, pp. 88-90.

Becker, A.; Morrison, H.F.; Zollinger, R. & Lazenby, P.G. (1986). Airborne Bathymetry and Sea-Bottom Profiling with the INPUT Airborne Electromagnetic System. In: *Airborne Resistivity Mapping, Geological Survey of Canada Paper 86-22*, G.J. Palacky (Ed.), pp. 107-109.

Bergeron, C.J., Jr.; Ioup, J.W. & Michel, G.A. (1989). Interpretation of Airborne Electromagnetic Data Using the Modified Image Method. *Geophysics*, Vol. 54, No. 8 (August 1989), pp. 1023-1030.

Bergeron, C.J. Jr.; Holladay, J.S.; Kovac, A.; Mozley, E.; Smith, B.D.; Watts, R.D. & Wright, D.L. (1987). Bathymetry and Related Uses Working Group Report. In: *Proceedings of the US Geological Survey Workshop on Development and Application of Modern Airborne Electromagnetic Surveys*, D. Fitterman (Ed.), October 7-9, 1987, pp. 204-205.

Boyd, G.W. (2004). HoistEM - A New Airborne Electriomagnetic System. *Conference Proceedings: PACRIM 2004*, ISBN: 9781920806187, Australian Institute of Mining and Metallurgy (September 2004), Adelaide, pp. 211-218.

Bryan, M.W.; Holladay, K.W.; Bergeron, C.J. Jr.; Ioup, J.W. & Ioup, G.E. (2003). MIM and Non-Linear Least-Squares Inversions of AEM Data in Barataria Basin, Louisiana. *Geophysics*, Vol. 68, No. 4 (July-August 2003), pp. 1126-1131.

Cox, L.H.; Wilson, G.A. & Zdanov, M.S. (2010). 3D Inversion of Airborne Electromagnetic Data Using a Moving Footprint. *Exploration Geophysics*, Vol. 41, No. 4 (December), pp 250-259.

Davis, A.C.; Macnae, J. & Robb, T. (2006). Pendulum Motion in Airborne HEM Systems. *Exploration Geophysics*, Vol. 37, No. 4 (December 2006), pp. 355-362.

Davis, A.; Macnae, J. & Hodges, G. (2009). Predictions of Bird Swing from GPS Coordinates. *Geophysics*, Vol. 74, No. 6 (November-December 2009), pp. F119-F126.

Dickinson, J.E.; Pool, D.R.; Groom, R.W. & Davis, L.J. (2010). Inference of Lithologic Distributions in an Alluvial Aquifer Using Airborne Transient Electromagnetic Surveys. *Geophysics*, Vol. 75, No. 4 (July-August 2010), pp. WA149-WA161.

Emerson,D.W. & Phipps, C.V.G. (1969). The Delineation of the Bedrock Configuration of Part of Port Jackson, New South Wales with a Boomer System. *Geophysical Prospecting*, Vol., 17, No. 3 (September 1969), pp. 219-230.

Fitterman, D.V. & Deszcz-Pan, M. (1998). Helicopter EM Mapping of Salt-Water Intrusion in Everglades National Park, Florida. *Exploration Geophysics*, Vol. 29, Nos 1 & 2, pp. 240-243.

Fountain, D. (1980). Airborne electromagnetic systems - 50 years of development. *Exploration Geophysics*, Vol. 29, Nos 1 & 2, pp. 1-11.

Fraser, D.C. (1978). Resistivity Mapping with an Airborne Multicoil Electromagnetic System. *Geophysics*, Vol. 43, No. 1 (February 1978), pp. 144-172.

Fullagar, P.K. & Oldenburg, D.W. (1984). Inversion of Horizontal Loop Electromagnetic Frequency Soundings. *Geophysics*, Vol. 49, No. 2 (February 1984), pp. 150-164.

Fullagar, P.K. (1989). Generation of Conductivity-Depth Pseudo-Sections from Coincident Loop and In-Loop TEM Data. *Exploration Geophysics*, Vol. 20, No., 1 & 2 (June, 1989), pp. 43-45.

Grant, F.S. & West, G.F. (1965). Interpretation Theory in Applied Geophysics. McGraw-Hill, New York.

Glover, P. (2009). What is the Cementation Exponent? A New Interpretation. *The leading Edge*, Vol. 28, No. 2 (January 2009), pp. 82-85.

Haas, C.; Gerland, S.; Eicken, H. & Miller, H. (1997). Comparison of Sea Ice Thickness Measurements Under Summer and Winter Conditions in the Arctic Using a Small Electromagnetic Induction Device. *Geophysics.*, Vol. 62, No. 3 (May-June 1997), pp. 749-757.

Haas, C.; Lobach, J.; Hendricks, S.; Rabenstein, L. & Pfaffling, A. (2009). Helicopter-borne Meaururements of Sea Ice Thickness Using a Small and Lightweight Digital EM System. *Journal of Applied Geophysics*, Vol. 67, No. 3 (March 2009), pp. 234-241.

Harris, G.A.; Vrbancich, J.; Keene, J. & Lean, J. (2001). Interpretation of Bedrock Topography Within the Port Jackson (Sydney Harbour) Region Using Marine Seismic Reflection. *15th Geophysical Conference of the Australian Society of Exploration Geophysicists*, Brisbane, Australia, 5-8 August 2001, *Extended Abstracts.*

Hatch, M.; Munday, T. & Heinson, G. (2010). A Comparative Study of In-river Geophysical Techniques to Define Variations in Riverbed Salt Load and Aid Managing River Salinization. *Geophysics*, Vol. 75, No. 4 (July-August 2010), pp. WA135-WA147.

Kirkegaard, C.; Sonnenborg, T.O.; Auken, E. & Jorgensen, F. (2011). Salinity Distribution in Heterogeneous Coastal Aquifers Mapped by Airborne Electromagnetics. *Vadoze Zone Journal*, Vol. 10, No. 1 (February 2011), pp. 125-135.

Kovacs, A. & Holladay, J.S. (1990). Sea-Ice Thickness Measurement Using a Small Airborne Sounding System. *Geophysics*, Vol. 55, No. 10 (October 1990), pp. 1327-1337.

Kovacs, A. & Valleau, C. (1987). Airborne Electromagnetic Measurement of Sea Ice Thickness and Sub-Ice Bathymetry. In: *Proceedings of the U.S. Geological Survey Workshop on Developments and Applications of Modern Airborne Electromagnetic Surveys*, D.V. Fitterman, (Ed.), 165-169, October 7-9, 1987.

Kovacs, A.; Holladay, J.S & Bergeron, C.J. Jr. (1995). The Footprint/Altitude Ratio for Helicopter Electromagnetic Sounding of Sea-Ice Thickness: Comparison of Theoretical and Field Estimates. *Geophysics*, Vol. 60, No. 2 (March-April 1995), pp. 374-380.

Kratzer, T. & Vrbancich, J. (2007). Real-Time Kinematic Tracking of Towed AEM Birds. *Exploration Geophysics*, Vol. 38, No. 2 (June 2010), pp. 132-143.

Ley-Cooper, Y.; Macnae, J.; Robb, T. & Vrbancich, J. (2006). Identification of Calibration Errors in Helicopter Electromagnetic (HEM) Data Through Transform to the Altitude-Corrected Phase-Amplitude Domain. *Geophysics*, Vol. 71, No. 2 (March-April 2006), pp. G27-G34.

Liu, G. & Becker, A. (1990). Two-Dimensional Mapping of Sea Ice Keels with Airborne Electromagnetics. *Geophysics*, Vol. 55, No. 2 (February, 1990), pp. 239-248.

Liu,G.; Kovacs, A. & Becker, A. (1991). Inversion of Airborne Electromagnetic SurveyData for Sea Ice Keel Shape. *Geophysics*, Vol. 56, No. 12 (December 1991), pp. 1986-1991.

Macnae, J.; Smith, R.; Polzer, B.D.; Lamontagne, Y. & Klinkert, P.S. (1991). Conductivity-Depth Imaging of Airborne Electromagnetic Step-Response Data. *Geophysics*, Vol. 56, No. 1 (January 1991), pp. 102-114.

Macnae, J.; King, A.; Stolz, N.; Osmakoff, A. & Blaha, A. (1998). Fast AEM Data Processing and Inversion. *Exploration Geophysics*, Vol. 29, No. 1 & 2 (June 1998), pp. 163-169.

Morrison, H.F.; Phillips, R.J. & O'Brien, D.P. (1969). Quantitative Interpretation of Transient Electromagnetic Fields Over a Layered Half Space. *Geophysical Prospecting*, Vol. 17, No. 1, (March 1969), pp. 82-101.

Morrison, H.F. & Becker, A. (1982). Analysis of Airborne Electromagnetic Systems for Mapping Depth of Seawater. Engineering Geoscience, University of California, Berkeley, California, Final report, ONR contract N00014-82-M-0073.

Mozley, E.C.; Kooney, T.N.; Byman, D.A & Fraley, D.E. (1991a). Kings Bay Airborne Electromagnetic Survey. Naval Oceanographic and Atmospheric Research

Laboratory, Marine Geosciences Division, Stennis Space Center, Report ID 019:352:91.

Mozley, E.C.; Kooney, T.N.; Byman, D.A & Fraley, D.E. (1991b). Airborne Electromagnetic Hydrographic Survey Technology. *SEG Expanded Abstracts*, Vol. 10, 468-471.

Nabighian, M.N. (1979). Quasi-static Transient Response of a Conducting Half-Space - an Approximate Representation. *Geophysics*, Vol. 44, No. 10 (October 1979), pp. 1700-1705.

Palacky, G.J. & West, G.F. (1991). Airborne Electromagnetic Methods. In *Electromagnetic Methods in Applied Geophysics – Volume 2 Applications Part B*, M.N. Nabighian, (Ed.), 811-879, Society of Exploration Geophysicists, ISBN 1-56080-22-4, Tulsa, Oklahoma.

Pelletier, R.E. & Holladay, K.W. (1994). Mapping Sediment and Water Properties in a Shallow Coastal Environment with Airborne Electromagnetic Profile Data: Case Study - the Cape Lookout, NC Area. *Marine Technology Society Journal*, Vol. 28, No. 2 (Summer 1994), pp. 57-67.

Pfaffling, A. & Reid, J.E. (2009). Sea Ice as an Evaluation Target for HEM Modelling and Inversion. *Journal of Applied Geophysics*, Vol. 67, No. 3 (March 2009), pp. 242-249.

Reid, J.E. & Macnae, J.C. (1998). Comments on the Electromagnetic "Smoke Ring" Concept. *Geophysics*, Vol. 63, No. 6, (November-December 1998), pp. 1908-1913.

Reid, J.E.; Vrbancich, J. & Worby, A.P. (2003a). A Comparison of Shipborne and Airborne Electromagnetic Methods for Antarctic Sea Ice Thickness Measurements. *Exploration Geophysics*, Vol. 34, Nos 1 & 2 (June 2003), pp. 46-50.

Reid, J.E.; Worby, A.P.; Vrbancich, J. & Munro, I.S. (2003b). Shipborne Electromagnetic Measurements of Antarctic Sea-Ice Thickness. *Geophysics*, Vol. 68, No. 5 (September-October 2003), pp. 1537-1546.

Reid, J.E. & Vrbancich, J. (2004). A Comparison of the Inductive-Limit Footprints of Airborne Electromagnetic Configurations. *Geophysics*, Vol. 69, No. 5 (September-October 2004), pp. 1229-1239.

Reid, J.E.; Pfaffling, A. & Vrbancich, J. (2006). Airborne Electromagnetic Footprints in 1D Earths. *Geophysics*, Vol. 71, No. 2 (March-April 2006), pp. G63-G72.

Sattel, D.; Lane, R.; Pears, G. & Vrbancich, J. (2004). Novel Ways to Process and Model GEOTEM data. *17th Geophysical Conference of the Australian Society of Exploration Geophysicists*, Sydney, Australia, 15-19 August 2004, *Extended Abstracts*.

Sattel, D. (2009). An Overview of Helicopter Time-Domain EM Systems. *20th Geophysical Conference of the Australian Society of Exploration Geophysicists*, Adelaide, Australia, 22-25 February 2009, *Extended Abstracts*.

Smith, R. (2001a). Tracking the Transmitting-Receiving Offset in Fixed-Wing Transient EM Systems: Methodology and Applications. *Exploration Geophysics*, Vol 32, No. 1 (March 2001), pp. 14-19.

Smith, R. (2001b). On Removing the Primary Field from Fixed-Wing Time-Domain Airborne Electromagnetic Data: Some Consequences for Quantitative Modelling, Estimating Bird Position and Detecting Perfect Conductors. *Geophysical Prospecting*, Vol. 49, No. 4 (July 2001), pp. 405-416.

Smith, R.S.; Hodges, R. & Lemieux, J. (2009). Case Histories Illustrating the Characteristics of the HeliGEOTEM System. *Exploration Geophysics*, Vol 40, No. 3 (September 2009), pp. 246-256.

Soininen, H.; Jokinen, T.; Oksama, M. & Suppala I. (1998). Sea Ice Thickness Mapping by Airborne and Ground EM Methods. *Exploration Geophysics*, Vol 29, Nos 1 & 2 (August 1998), pp. 244-248.

Son, K.H. (1985). Interpretation of Electromagnetic Dipole-Dipole Frequency Sounding Data Over a Vertically Stratified Earth. PhD thesis, North Carolina State University, Raleigh, 149 pp.

Sorensen, K.I. & Auken, E. (2004). SkyTEM – A New High-Resolution Helicopter Transient Electromagnetic System. *Exploration Geophysics*, Vol. 35, No. 3 (September 2004), pp. 194-202.

Spies, B.; Fitterman, D.; Holladay, S. & Liu, G. (Editors) (1998). Proceedings of the International Conference on Airborne Electromagnetics (AEM 98). *Exploration Geophysics* 1998, Vol. 29, Nos 1 and 2 (August 1998), pp. 1-262.

Vrbancich, J.; Hallett, M. & Hodges, G. (2000a). Airborne Electromagnetic Bathymetry of Sydney Harbour. *Exploration Geophysics*, Vol. 31, No. 1&2 (June 2000), pp. 179-186.

Vrbancich, J.; Fullagar, P.K. & Macnae, J. (2000b). Bathymetry and Seafloor Mapping via One Dimensional Inversion and Conductivity Depth Imaging of AEM. *Exploration Geophysics*, Vol. 31, No. 4 (December 2000), pp. 603-6.

Vrbancich, J. (2004). Airborne Electromagnetic Bathymetry Methods for Mapping Shallow Water Sea Depths. *International Hydrographic Review*, Vol. 5, No. 3 (November 2004), pp. 59-84.

Vrbancich, J. & Smith, R. (2005). Limitations of Two Approximate Methods for Determining the AEM Bird Position in a Conductive Environment. *Exploration Geophysics*, Vol. 36, No. 4 (December 2005), pp. 365-373.

Vrbancich, J.; Sattel, D.; Annetts, D.; Macnae, J. & Lane, R. (2005a). A Case Study of AEM Bathymetry in Geographe Bay and Over Cape Naturaliste, Western Australia, Part 1: 25 Hz QUESTEM. *Exploration Geophysics*, Vol. 36, No. 3 (September 2005), pp. 301-309.

Vrbancich, J.; Macnae, J.; Sattel, D. & Wolfgram, P. (2005b). A Case Study of AEM Bathymetry in Geographe Bay and Over Cape Naturaliste, Western Australia, Part 2: 25 and 12.5 Hz GEOTEM. *Exploration Geophysics*, Vol. 36, No. 4 (December 2005), pp. 381-392.

Vrbancich, J. & Fullagar, P.K. (2004). Towards Seawater Depth Determination Using the Helicopter HoistEM System. *Exploration Geophysics*, Vol. 35, No. 4 (December 2004), pp. 292-296.

Vrbancich, J. & Fullagar, P.K. (2007a). Towards Remote Sensing of Sediment Thickness and Depth to Bedrock in Shallow Seawater Using Airborne TEM. *Exploration Geophysics*, Vol. 38, No. 1 (March 2007), pp. 77-88.

Vrbancich, J. & Fullagar, P.K. (2007b). Improved Seawater Depth Determination Using Corrected Helicopter Time Domain Electromagnetic Data. *Geophysical Prospecting*, Vol. 55, No. 3 (May 2007), pp. 407-420.

Vrbancich, J. (2009). An Investigation of Seawater and Sediment Depth Using a Prototype Airborne Electromagnetic Instrumentation System – A Case Study in Broken Bay, Australia. *Geophysical Prospecting*, Vol. 57, No. 4 (July 2007), pp. 633-651.

Vrbancich, J.; Whiteley, R.J. & Emerson, D.W. (2011a). Marine Seismic Profiling and Shallow Marine Sand Resistivity Investigations in Jervis Bay, NSW, Australia. *Exploration Geophysics*, Vol. 42, No. 2 (June 2011), pp. 127-138.

Vrbancich, J.; Whiteley, R.J.; Caffi, P. & Emerson, D.W. (2011b). Marine Seismic Profiling and Shallow Marine Sand Resistivity Investigations in Broken Bay, NSW, Australia. *Exploration Geophysics*, Vol. 42, No. 4 (December), 227-238, DOI 10.1071/EG11031.

Vrbancich, J.; Fullagar, P. & Smith, R. (2010). Testing the Limits of AEM Bathymetry with a Floating TEM System. *Geophysics*, Vol. 75, No. 4 (July-August 2010), pp. WA163-WA177.

Vrbancich, J. (2010). Preliminary Investigations Using a Helicopter Time-Domain System for Bathymetric Measurements and Depth-to-Bedrock Estimation in Shallow Coastal Waters - A Case Sudy in Broken Bay, Australia, *Proceedings OCEANS 2010 IEEE*, ISBN 9781424452217, Sydney, Australia, May 24-27, 2010, 9pp.

Vrbancich, J. (2011). AEM Applied to Bathymetric Investigations in Port Lincoln, South Australia - Comparison with an Equivalent Floating TEM System. *Exploration Geophysics*, Vol. 42, No. 3 (September, 2011), pp. 167-175.

Wait, J.R. (1982). *Geo-Electromagnetism*. Academic Press, ISBN 0-12-730880-6, New York.

Ward, S.H. & Hohman, G.W. (1987). Electromagnetic Theory for Geophysical Applications. In *Electromagnetic Methods in Applied Geophysics - Volume 1 Theory*, M.N. Nabighian, (Ed.), 131-311, Society of Exploration Geophysicists, ISBN 0-931830-51-6, Tulsa, Oklahoma.

Weaver, J.T. (1994). *Mathematical Methods for Geo-Electromagnetic Induction*. Research Studies Press Ltd., ISBN 0 86380 165 X, Taunton, Somerset, England.

West, G.F. & Macnae, J.C. (1991). Physics of the Electromagnetic Induction Exploration Method. In *Electromagnetic Methods in Applied Geophysics - Volume 2 Applications Part A*, M.N. Nabighian, (Ed.), 5-45, Society of Exploration Geophysicists, ISBN 1-56080-22-4, Tulsa, Oklahoma.

Witherly, K. (2004). The Geotech VTEM Time Domain Helicopter EM System. *SEG Expanded Abstracts*, Vol. 23, Section MIN 3.5 (October 2004), 1221-1224.

Wolfgram, P. & Karlik, G. (1995). Conductivity-Depth Transform of GEOTEM Data. *Exploration Geophysics*, Vol. 26, No. 2 & 3 (September 1995), pp. 179-185.

Wolfgram, P. & Vrbancich, J. (2007). Layered Earth Inversions of AEM Bathymetry Data Incorporating Aircraft Attitude and Bird Offset - A Case Study of Torres Strait. *Exploration Geophysics*, Vol. 38, No. 2 (June 2007), pp. 144-149.

Won, I.J. & Smits, K. (1985). Airborne Electromagnetic Bathymetry: Naval Ocean and Research Development Activity, NSTL, MS, NORDA Report 94.

Won, I.J. & Smits, K. (1986a). Characterization of Shallow Ocean Sediments Using the Airborne Electromagnetic Method. *IEEE Journal of Oceanic Engineering*, Vol. OE-11, No. 1 (January 1986), pp. 113-122.

Won, I.J.,& Smits, K. (1986b). Application of the Airborne Electromagnetic Method for Bathymetric Charting in Shallow Oceans. In: *Airborne Resistivity Mapping, Geological Survey of Canada Paper 86-22*, G.J. Palacky (Ed.), pp. 99-106.

Won, I.J. & Smits, K. (1987a). Airborne Electromagnetic Bathymetry. In: *Proceedings of the US Geological Survey Workshop on Development and Application of Modern Airborne Electromagnetic Surveys*, D. Fitterman (Ed.), October 7-9, 1987, pp. 155-164.

Won, I.J. & Smits, K. (1987b). Airborne Electromagnetic Measurements of Electrical Conductivity of Seawater and Bottom Sediments Over Shallow Ocean. *Marine Geotechnology*, Vol. 7, No. 1, (March 1987), pp. 1-14.

Zollinger, R.D. (1985). Airborne Electromagnetic Bathymetry. MSc thesis, University of California, Berkeley, California, 45 pp.

Zollinger, R.; Morrison, H.F.; Lazenby, P.G. & Becker, A. (1987). Airborne Electromagnetic Bathymetry. *Geophysics*, Vol. 52, No. 8 (August 1987), pp. 1127-1137.

3

3-D Coastal Bathymetry Simulation from Airborne TOPSAR Polarized Data

Maged Marghany
Institute for Science and Technology Geospatial (INSTEG),
Universiti Teknologi Malaysia, Skudai, Johore Bahru
Malaysia

1. Introduction

Remote sensing techniques in real time could be a major tool for bathymetry mapping which could produce synoptic data overlarge areas at extremely low cost. In contrast, conventional techniques such as single- or multi-beam ship-borne echoes are costly and time-consuming, especially when large areas are being surveyed (Marghany et al., 2009b). The ocean bathymetry features can image by radar in coastal waters with strong tidal currents (Vogelzang et al., 1992; Vogelzang et al., 1997; Hesselmans et al., 2000, Marghany et al., 2010). Under practical circumstances, synthetic aperture radar (SAR) is able to detect shallow ocean bathymetry features (Alpers and Hennings 1984; Shuchman et al. 1985; and Vogelzang 1997). According to Alpers and Henning (1984) ocean bathymetry can determine indirectly based on means of sea surface variations that caused by the gradient current overflowing the submarine features. Therefore, this concept is valid with the presence of strong current associates with capillary waves. Under this circumstance, SAR antenna receives a strong backscatter from ocean surface. Nevertheless, multi-beam ship-borne echoes provide bathymetry to water depths above 100 m whereas SAR data are limited to less than 25 m. Under local circumstances such as strong tidal gradient and wind speed higher than 3 m/s, SAR data can detect shallow ocean bathymetry features down to 20 m (Marghany et al., 2009b).

1.1 Principle of SAR ocean bathymetry imaging

Several theories concerning the radar imaging mechanism of underwater bathymetry have been established, such as by Alpers and Hennings (1984); Shuchman et al. (1985); and Vogelzang (1997). The physical theories describing the radar imaging mechanisms for ocean bathymetry are well understood as three stages: (i) the modulation of the current by the underwater features, (ii) the modulation of the sea surface waves by the variable surface current, and (iii) the interaction of the microwaves with the surface waves (Alpers and Hennings, 1984) (Fig. 1). The imaging mechanism which reflects sea bottom topography in a given SAR image consists of three models. These models are a flow model, a wave model and the SAR backscatter model. These theories are the basis of commercial services which generate bathymetric charts by inverting SAR images at a significantly lower cost than conventional survey techniques (Wensink and Campbell, 1997). In this context, Hesselmans et al. (2000) developed the Bathymetry Assessment System, a computer program which can be used to calculate the depth from any SAR image and a limited number of sounding data points. They found that the imaging model was suitable for simulating a SAR image from the depth map.

It showed good agreement between the backscatter in both the simulated and airborne-acquired images, when compared, with accuracy (root mean square) error of $\pm$ 0.23 m of order of 10 m within a coastal bathymetry range of 25-30 m. Recently, Li et al., (2009) are utilized RADARSAT-1 and ENVISAT synthetic aperture radar (SAR) images for mapping sand ridges with 30 m water depth. In doing so, they used modelled tidal current as to an advanced radar-imaging model to simulate the SAR image at a given satellite look angle and for various types of bathymetry. In this regard, the shallow-water bathymetry is acquired in a 2-D space. Finally, they reported that the sand ridge can be imaged when strong ocean currents exist. On the contrary, Lyzenga et al., (2006) used a simple method of estimating water depths from multispectral imagery, based on an approximate shallow-water reflectance model. They found that a single set of coefficients derived from a set of IKONOS images produces the good performance with an aggregate RMS error of $\pm$ 2.3 m which is higher than bathymetry retrieved from SAR (Hesselmans et al. 2000). Coastal bathymetry mapping by using optical remote sensing data, however, can be only fully utilized in the clearest water, and considerably less in turbid water (Vogelzang et al., 1992). Indeed, as the different wavelengths pass through the water column they become attenuated by the interaction with suspended particles in water (Mills, 2008).

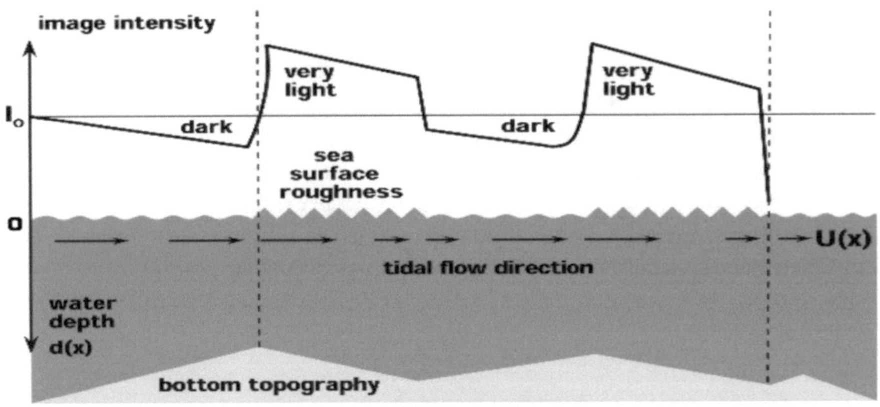

Fig. 1. SAR 's concept for imaging ocean bathymetry ((Alpers and Hennings, 1984).

1.2 Speckle impact on SAR ocean bathymetry imaging

The high speckle noise in SAR images has posed great difficulties in inverting SAR images for simulating coastal bathymetry. Speckle is a result of coherent interference effects among scatterers which are randomly scattered within each resolution cell. The speckle size, which is a function of the spatial resolution, induces errors in bathymetry signature detection. Reducing the speckle effects, require appropriate filters, i.e. Lee, Gaussian, etc. (Lee et al 2002, Marghany and Mazlan 2010), could be used in the pre-processing stage. The effectiveness of these speckle-reducing filters is however much influenced by cause's factors and application. Since the SAR images the sea surface, all speckles in SAR images are to function of local changes in the surface roughness because, direct reduction of the wave height (because of slicks), wind impelled roughness changes (atmospheric effects) or wave-current interactions (fronts and bathymetry).

In contrast, Yu and Scott (2002) stated several restrictions of the speckle filtering approach. They reported size and shape of the filter window can affect the accuracy of despeckle filters. For instance, extremely large window size will form a blurred image, while a small window will decrease the smoothing competence of the filter and will leave speckle. They also found that window size can change the physical characteristics of targets in SAR image. For instance a square window (as is typically applied) will lead to corner rounding of rectangular features. Despite using despeckle filters to perform edge enhancement, speckle in the neighborhood of an edge (or in the neighborhood of a point feature with high contrast) will remain after filtering. Additionally, the thresholds used in the enhanced filters, although motivated by statistical arguments, are *ad hoc* improvements that only demonstrate the insufficiency of the window-based approaches. The hard thresholds that enact neighborhood averaging and identity filtering in the extreme cases lead to blotching artifacts from averaging filtering and noisy boundaries from leaving the sharp features unfiltered (Yu and Scott 2002).

In the case of bathymetry mapping not all the filters stated in the literature are suitable for bathymetry application. In fact, SAR data have discontinuities and lower grey levels gradient (Marghany et al., 2009a). Besides, by applying some kinds of filter such as Lee, most of bathymetry signature information will be lost. In this perspective, Inglada and Garello, (1999) and Marghany et al., (2009b) stated that an anisotropic diffusion filter is more appropriate for speckle reduction in the case of bathymetry signature detection in a SAR image. They concluded that it produced the highest smoothed image as the anisotropic diffusion filter preserves the mean grey-level and maintains the bathymetry signature compared to Lee filter. Nevertheless, Inglada and Garello, (1999 and 2002) were not able to state the accuracy rate of utilizing the Volterra model (Section 4.2) and anisotropic diffusion filter for SAR.

1.3 Hypothesis of study

Concerning with above prospective, we address the question of despeckles' impact on the accuracy of retrieving ocean bathymetry without needing to include any sounding data values. This was demonstrated with airborne SAR data (namely the TOPSAR) using integration of the Volterra kernel (Inglada and Garello, 1999) and fuzzy B-spline algorithm (Marghany and Mazlan 2005 and Marghany et al., 2007). Nonetheless, the studies of Marghany and Mazlan (2006) and Marghany et al., (2007) have failed to derive accurate bathymetry depth with single Cvv band although the root mean square error is ± 9 m. Five hypotheses are examined:

- the Volterra model can use to detect tidal current pattern from TOPSAR polarised data,
- anisotropic diffusion algorithm can reduce the speckle in SAR data and determine sharp bathymetry feature;
- there are significant differences between the different bands in detecting ocean currents,
- the continuity equation can be used to obtain the water depth, and
- fuzzy B-splines can be implemented to invert the water depth values determined by the continuity equation into 3-D bathymetry.

2. Study area

The study area is located in the coast of Kuala Terengganu, on eastern part of Malaysia Peninsula. This area is approximately 20 km along the north of Kuala Terengganu coastline, located between 5^0 21′ N , 103^0 10′ E and 5^0 30′ N, 103^0 20′ E). Sand materials make up the entire of this beach (Marghany 1994). The east coast of Peninsular Malaysia is annually subjected to the northeast monsoon wind (November to January) (Marghany et al., 2010) showed that the mean, and longer significant wave periods were 8 to 10 seconds.

Significant wave height maximum were reported as 4 m and 2.4 m, respectively in February and March. However, during the south-west monsoon wave height was ranged between 0.4m -0.7m (Marghany et al., 2010). During the inter- monsoon period (September to mid of November), the wave height was ranged between 0.37 m to 1.6 m (Marghany et al., 2010). According to Marghany et al., (2010) the coastal water less than 50 nautical miles from shore is quite shallow with the deepest area being approximately 50 m (Fig. 2). The bottom has gentle slopes, gradually deepening towards the open sea. A clear feature of this area is the primary hydrologic communications between the estuary and the South China Sea which is the largest estuary along the Kuala Terengganu (Marghany et al., 2009b).

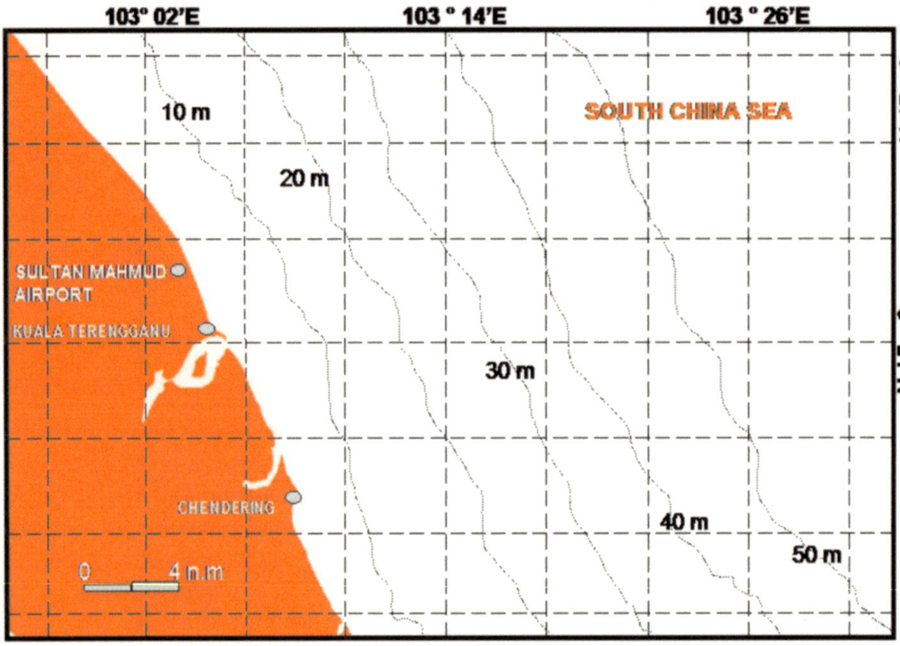

Fig. 2. Bathymetry along the coastal waters of Kuala Terengganu.

3. Data sets

Airborne data acquired in this study were derived from the Jet Propulsion Laboratory (JPL) airborne Topographic Synthetic Aperture Radar (TOPSAR) data on December 6 ,1996. TOPSAR is a NASA/JPL multi-frequency radar imaging system aboard a DC-8 aircraft and operated by NASA's Ames Research Center at Moffett Field, USA. TOPSAR data are fully polarimetric SAR data acquired with HH-,VV-, HV- and VH-polarized signals from 5 m x 5 m pixels, recorded for three wavelengths: C band (5 cm), L band (24 cm) and P band (68 cm). The full set of C-band and L-band have linear polarizations (HH, VV, HV), phase differences (HHVV), and circular polarizations (RR, RL).

In addition, the TOPSAR sensor uses two antennas to receive the radar backscatter from the surface. The difference in arrival times of the return signals at the two antennas is converted into a modulo -2π phase difference. Further, TOPSAR data with C-band provides digital elevation model with rms error in elevation ranging from about 1 m in the near range to greater than 3 m in the far range. A further explanation of TOPSAR data acquisition is given by Melba et al. (1999). This study utilizes both C_{VV} and L_{HH} bands for 3-D bathymetry reconstruction because of the widely known facts of the good interaction of VV and HH polarization to oceanographic physical elements such as ocean wave, surface current features, etc. Elaboration of such further explanation can be found in (Alpers and Hennings 1984; and Inglada and Garello 2002; Marghany et al., 2010).

4. Model for 3-D bathymetry retrieving

Three models involved for 3-D bathymetry retrieving from TOPSAR polarized data: anisotropic diffusion, algorithm the Volterra model and the fuzzy B-spline model. The Volterra model is used to assimilate the tidal current flow from TOPSAR data. The simulation current velocity aims to retrieve water depth gradients under tidal current flow spatial variations. The fuzzy B-spline used to remodel the three-dimensional (3-D) water depth from a 2-D continuity equation.

4.1 Anisotropic diffusion algorithm

Anisotropic diffusion is technique that aims at reducing image noise while preserving edges, lines and other details that are important for the image interpretation (Perona and Malik 1990). Formally, let $\Omega \subset \mathbb{R}^2$ denote a subset of the TOPSAR image plane $I_{TOPSAR}(.,t): \Omega \rightarrow \mathbb{R}$ which is part of TOPSAR intensity. Therefore, anisotropic diffusion is given by

$$\begin{aligned}\frac{\partial I_{TOPSAR}}{\partial t} &= div(c(i,j,t)\Delta I_{TOPSAR}) \\ &= \nabla c.\nabla I_{TOPSAR} + c(i,j,t)\Delta I_{TOPSAR}\end{aligned} \tag{1}$$

where, $c(i,j,t)$ is the diffusion coefficient and $c(i,j,t)$ controls the rate of diffusion that is usually chosen as a function of the TOPSAR image intensity gradient ΔI_{TOPSAR} to preserve bathymetry edge in TOPSAR data. Besides, Δ denotes the Laplacian, ∇ denotes the gradient, and *div (.......)* is the divergence operator. Following Perona and Malik (1987), two functions for the diffusion coefficient are considered:

$$c(\|\nabla I_{TOPSAR}\| = e^{-(\|\nabla I_{TOPSAR}\|/K)^2} \tag{2}$$

and

$$c(\|\nabla I_{TOPSAR}\| = (1 + \|\nabla I_{TOPSAR}\|)^{-1} \tag{3}$$

the constant K controls the sensitivity to edges and is usually chosen as a function of the speckle variations in TOPSAR data. In this regard, $c(\|\nabla I_{TOPSAR}\| \to 0$ if $\|\nabla I_{TOPSAR}\| \gg K$ that leads to all-pass filter. Besides isotropic diffusion (Gaussian filtering achieves under boundary condition of $c(\|\nabla I_{TOPSAR}\| \to 1$ when $\|\nabla I_{TOPSAR}\| \ll K$. Anisotropic diffusion resembles the process that creates a scale-space, where an image generates a parameterized family of successively more and more blurred images based on a diffusion process (Sapiro 2001).

4.2 Volterra algorithm

According to Marghany et al., (2010), the Volterra algorithm can use to express the SAR image intensity as a series of nonlinear filters on the ocean surface current. This means the Volterra algorithm can use to study the image energy variation as a function of parameters such as the current direction, or the current magnitude. A generalized, nonparametric framework to describe the input-output x and y signals relation of a time-invariant nonlinear system is provided by Inglada and Garello (1999). Additional, the input x corresponds to the different TOPSAR band intensities i.e., (C and L bands) whereas y corresponds to Volterra series of the different bands. In discrete form, the Volterra series for input, TOPSAR data intensities X (n), and output of TOPSAR signals in form of Volterra series, Y (n) as given by Inglada and Garello (2002) can be expressed as:

$$\begin{aligned} Y(n) &= h_0 + \sum_{i_1=1}^{\infty} h_1(i_1)X(n-i_1) + \sum_{i_1=1}^{\infty}\sum_{i_2=1}^{\infty} h_2(i_1,i_2)X(n-i_1)X(n-i_2) + \\ &\sum_{i_1=1}^{\infty}\sum_{i_2=1}^{\infty}\sum_{i_3=1}^{\infty} h_3(i_1,i_2,i_3)X(n-i_1)X(n-i_2)X(n-i_3) + \dots\dots\dots + \\ &\sum_{i_1=1}^{\infty}\sum_{i_2=1}^{\infty}\dots\dots\dots\dots\sum_{i_k=1}^{\infty} h_k(i_1,i_2,\dots\dots,i_k)X(n-i_1)X(n-i_2)\dots\dots\dots\dots X(n-i_k) \end{aligned} \tag{4}$$

where, n, i_1, i_2,...,i_k are discrete time lags. The function h_k (i_1 ,i_2 ,...,i_k) is the kth-order Volterra kernel characterizing the system. h_1 is the kernel of the first order Volterra functional, which performs a linear operation on the input and h_2, h_3,...,h_k capture the nonlinear interactions between input and output TOPSAR signals. In this context, the nonlinearity is expressed as the relationship between different TOPSAR band intensities and ocean surface roughness. Consequently, surface current gradients in shallow waters can be imaged by TOPSAR different bands through energy transfer towards the waves. Indeed, the radar system is restricted to measure surface roughness. The order of the non-linearity is the highest effective order of the multiple summations in the functional series (Marghany et al., 2009b).

Following Marghany and Mazlan (2006); Marghany et al., (2009b) and Marghany et al., (2010) Fourier transform is used to acquire nonlinearity function from equation 4 as given by

$$Y(v) = FT[Y(n)] = \int Y(n)e^{-j2\pi vn}dn \tag{5}$$

where, v is frequency and $j = \sqrt{-1}$ (Marghany et al., 2010). Domain frequency of TOPSAR image $I_{TOPSAR}(v,\Psi)$ can be described by using equation 5 with following expression

$$I_{TOPSAR}(v,\Psi_0) = FT[I(r,a)e^{j(R/V)u_d(r,a)}] \tag{6}$$

where $I(r,a)$ is the intensity TOPSAR image pixel of azimuth *(a)* and range *(r)*, respectively, ψ_0 is the wave spectra energy and *R/V* is the range to platform velocity ratio, in case of TOPSAR equals 32 s and $U_d(r,a)$ is the radial component of surface velocities (Inglada and Garello 2002). Marghany et al., (2010) stated that equation 6 does not satisfy the relationship between TOPSAR data and ocean surface roughness. More precisely, the action balance equation (ABE) describes the relationship between surface velocity $\vec{u}$, and its gradient and the action spectral density ψ of the short surface wave i.e. Bragg wave (Alpers and Hennings 1984). In reference to Inglada and Garello (1999), the expression of ABE into first-order Volterra kernel $H_1(v_a,v_r)$ of frequency domain for the current flow in the range direction can be described as:

$$H_1(v_a,v_r) = k_r\langle\vec{U}\rangle\vec{v}\left[\vec{K}^{-1}[\frac{\partial\psi}{\partial\vec{k}} + \frac{\partial\vec{c}_g}{\partial x_a}\vec{v} + \frac{\partial\overline{u}_r}{\partial y_r}v_r + 0.5u_a]\right]\left[\frac{\partial\psi}{\partial\omega}\right]$$
$$\frac{\vec{c}_g(\vec{K})\vec{v} + j.0.5\omega_0^{-1}}{[\vec{c}_g(\vec{K})\vec{v}]^2 + [0.5\omega_0^{-1}]^2} + j(6.10^{-3}.\vec{K}^{-4})(\frac{R}{V})v_r \tag{7}$$

where, $\langle\overrightarrow{U}\rangle$ is the mean current velocity, $\vec{u}_r$ is the current flow along the range direction while $\vec{u}_a$ is current gradient along the azimuth direction. K_r is the wave number along the range direction, $\vec{K}$ is the spectra wave vector, ω_0 is the angular wave frequency, $\vec{c}_g$ is the group velocity, ψ is the wave spectra energy, v stands for the Volterra kernel frequency along the azimuth and range directions and *R/V* is the range to platform velocity ratio.

Then, the domain frequency of TOPSAR data $I_{TOPSAR}(v,\Psi)$ can be expressed by using Volterra model for ABE into equation 6

$$I_{TOPSARV}(v,\Psi) = FFT[(\Psi_0(a,r) + \int Y(n)).\sum_{N=0}^{+\infty}\frac{1}{n!}(j\frac{R}{V}u_r(a,r))^N] \tag{8}$$

where $N = 1,2,3,..........k$ and $I_{TOPSARV}(v,\Psi)$ represents Volterra kernels for the TOPSAR image in frequency domain in which can be used to estimate mean current flow $\vec{U}_r(0,r)$ in the range direction (*r*) with the following expression (Inglada and Garello, 2002)

$$I_{TOPSARV} = \vec{U}_r(0,r).H_{1r}(v_a,v_r) \tag{9}$$

The mean current movement along the range direction can be calculated by using the formula was proposed by Vogelzang et al. (1997)

$$\vec{U}_r(v_a,v_r) = \frac{FFT\left[\prod_{j=1}^{i} I_{TOPSARV}(t)\right]}{H_{1r}(v_a,v_r)} \tag{10}$$

where $FFT\left[\prod_{j=1}^{i} I_{TOPSARV}(t)\right]$ is the linearity of the Fourier transform for the input TOPSAR image intensity $I_{TOPSARV}(t)$ i.e. t is time domain. The inverse filter $P(v_a,v_r)$ is used since $H_{1r}(v_a,v_r)$ has a zero for $\vec{U}_r(v_a,v_r)$ which indicates that the mean current velocity should have a constant offset. The inverse filter $P(v_a,v_r)$ can be given as

$$P(v_a,v_r)=\begin{cases}[H_{1r}(v_a,v_r)]^{-1} & if(v_a,v_r)\neq 0,\\ 0 & \text{Otherwise.}\end{cases} \quad (11)$$

Then, the continuity equation is used to estimate the water depth as given by Vogelzang et al. (1992)

$$\frac{\partial \zeta}{\partial t}+\nabla.\{(d+\zeta)\vec{U}_r(0,v_r)\}=0 \quad (12)$$

where ζ is the surface elevation above the mean sea level, which is obtained from the tidal table, t is the time and d is the local water depth. The real current data was estimated from the Malaysian tidal table of 6 December, 1996 (Malaysian Department of Survey and Mapping 1996).

4.3 The fuzzy B-splines method

The fuzzy B-splines (FBS) are introduced allowing fuzzy numbers instead of intervals in the definition of the B-splines. Typically, in computer graphics, two objective quality definitions for fuzzy B-splines are used: triangle-based criteria and edge-based criteria (Marghany et al., 2009a). A fuzzy number is defined using interval analysis. There are two basic notions that we combine together: confidence interval and presumption level. A confidence interval is a real values interval which provides the sharpest enclosing range for current gradient values.

An assumption μ -level is an estimated truth value in the [0, 1] interval on our knowledge level of the gradient current (Anile 1997). The 0 value corresponds to minimum knowledge of gradient current, and 1 to the maximum gradient current. A fuzzy number is then prearranged in the confidence interval set, each one related to an assumption level $\mu \in$ [0, 1]. Moreover, the following must hold for each pair of confidence intervals which define a number: $\mu \succ \mu' \Rightarrow d \succ d'$.

Let us consider a function $f: d \rightarrow d'$, of N fuzzy variables $d_1, d_2, ..., d_n$. Where d_n are the global minimum and maximum values of the water depth of the function on the current gradient along the space. Based on the spatial variation of the gradient current, and water depth, the fuzzy B-spline algorithm is used to compute the function f (Marghany et al., 2010).

Marghany et al., (2010) assumed that $d(i,j)$ is the depth value at location i,j in the region D where i is the horizontal and j is the vertical coordinates of a grid of m times n rectangular cells. Let N be the set of eight neighbouring cells. The input variables of the fuzzy are the amplitude differences of water depth d defined by (Anile et al. 1997):

$$\Delta d_N = d_i - d_0, N = 1,........,8 \quad (13)$$

where the d_i, N=1, 8 values are the neighbouring cells of the actually processed cell d_0 along the horizontal coordinate *i*. To estimate the fuzzy number of water depth d_j which is located along the vertical coordinate *j*, we estimated the membership function values μ and $\mu^{'}$ of the fuzzy variables d_i and d_j, respectively by the following equations were described by Rövid et al. (2004)

$$\mu = \max\left\{\min\left\{m_{pl}(\Delta d_i) : d_i \in N_i\right\}; N = 1....,9\right\} \quad (14)$$

$$\mu^{'} = \max\left\{\min\left\{m_{LNl}(\Delta d_i) : d_i \in N_i\right\}; N = 1....,9\right\} \quad (15)$$

where m_{pl} and m_{LNl} correspond to the membership functions of fuzzy sets. From equations 11 and 12, one can estimate the fuzzy number of water depth d_j

$$d_j = d_i + (L-1)\Delta\mu \quad (16)$$

where $\Delta\mu$ is $\mu - \mu^{'}$ and $L = \{d_1,............,d_N\}$. Equations 15 and 16 represent water depth in 2-D, in order to reconstruct fuzzy values of water depth in 3-D, then fuzzy number of water depth in *z* coordinate is estimated by the following equation proposed by Russo (1998) and Marghany et al., (2010),

$$d_z = \Delta\mu MAX\{m_{LA}\left|d_{i-1,j} - d_{i,j}\right|, m_{LA}\left|d_{i,j-1} - d_{i,j}\right|\} \quad (17)$$

where d_z fuzzy set of water depth values in z coordinate which is function of *i* and *j* coordinates i.e. $d_z = F(d_i, d_j)$. Fuzzy number F_O for water depth in *i,j* and *z* coordinates then can be given by

$$F_O = \{\min(d_{z_0},............,d_{z_\Omega}), \max(d_{z_0},............,d_{z_\Omega})\} \quad (18)$$

where Ω =1, 2, 3, 4,

The fuzzy number of water depth F_O then is defined by B-spline in order to reconstruct 3-D of water depth. In doing so, B-spline functions including the knot positions, and fuzzy set of control points are constructed. The requirements for B-spline surface are set of control points, set of weights and three sets of knot vectors and are parameterized in the *p* and *q* directions.

Following Marghany et al., (2009b) and Marghany et al., (2010), as in the Volterra algorithm, data are derived from the TOPSAR polarised backscatter images by the application of a 2-dimensional fast Fourier transform (2DFFT). First, each estimated current data value in a fixed kernel window size of 512 x 512 pixels and lines is considered as a triangular fuzzy number defined by a minimum, maximum and measured value. Among all the fuzzy numbers falling within a kernel window size, a fuzzy number is defined whose range is given by the minimum and maximum values of gradient current and water depth along each kernel window size. Furthermore, the identification of a fuzzy number is acquired to summarize the estimated water depth data in a cell and it is characterized by a suitable membership function. The choice of the most appropriate membership is based on triangular numbers which are identified by minimum, maximum, and mean values of water depth estimated by continuity equation. Furthermore, the membership support is the range of water depth data in the cell and whose vertex is the median value of water depth data (Anile et al. 1997).

5. Three-dimensional ocean bathymetry from TOPSAR data

Figure 3 shows the signature of the underwater topography. The signature of underwater topography is obvious as frontal lines parallel to the shoreline. The backscattered intensity is damped by -2 to -10 dB compared to the surrounding water environment in L -band with HH polarization and -6 to -14 dB in C-band with VV polarization data (Fig. 3).

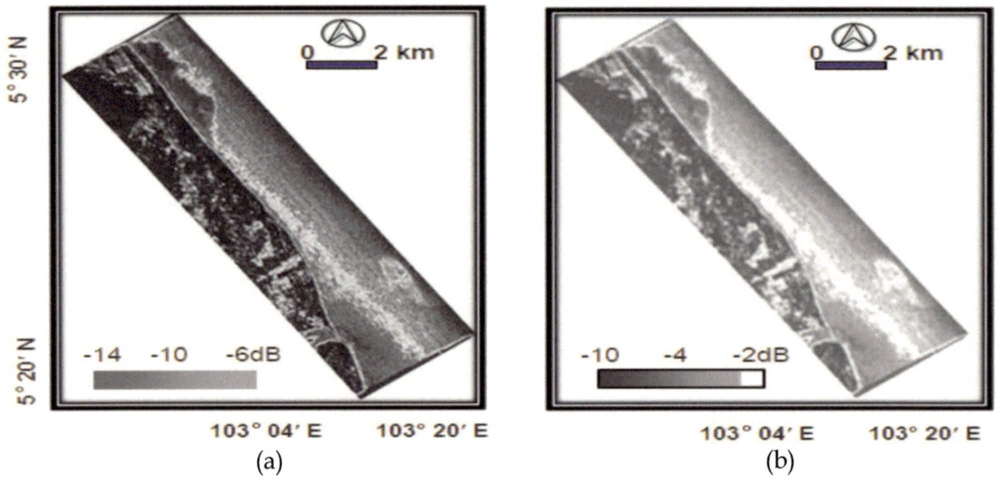

Fig. 3. Bathymetry Signature with Different Bands of (a) C_{VV} and (b) L_{HH} bands.

Figure 4 shows the clearer bathymetry signature is extracted by utilizing anisotropic diffusion algorithm. L_{HH} band has clear bathymetry feature than C_{VV} band. In fact, anisotropic diffusion algorithm is able to extract boundary edge for both horizontal and vertical direction. Otherwise said, it is synthesized boundary edge (Maeda et al. 1997 and Marghany et al., 2009b). In this regard, anisotropic diffusion resembles the process that creates a scale-space, where an image generates a parameterized family of successively more and more blurred images based on a diffusion process. Each of the resulting images in this family are given as a convolution between the image and a 2-D isotropic Gaussian filter, where the width of the filter increases with the parameter (Fig. 4). This diffusion process is a linear and space-invariant transformation of the original image. Anisotropic diffusion is a generalization of this diffusion process: it produces a family of parameterized images, but each resulting image is a combination between the original image and a filter that depends on the local content of the original image. As a consequence, anisotropic diffusion is a non-linear and space-variant transformation of the original image.

In addition, both C_{vv} and L_{HH} bands show bathymetry signature is concided with water depths which are ranged between 5 m to 20 m (Fig.4). The results show the potential of TOPSAR data for ocean bathymetry reconstruction where TOPSAR L_{HH} band backscatter across bathymetry signature pixels agrees satisfactorily with previous published results

(Vogelzang et al. 1992; Inglada and Garello 1999; Marghany et al., 2007). This is due to the fact that the ocean signature of the boundary is clear in the brightness of a radar return, since the backscatter tends to be proportional to wave height (Vogelzang et al. 1992). In C-band with VV polarization, this feature is clearly weaker than at L-band with HH polarization. In fact, L_{HH} band has higher backscatter value of 2 dB than C_{VV} band. In this context, it is possible that the character of the current gradient is such that the L_{HH} band surface Bragg waves are more strongly modulated than for C_{VV} band. This may provide an explanation for weaker bathymetric signatures at C_{VV} band. The finding is similar to that of Romeiser and Alpers (1997).

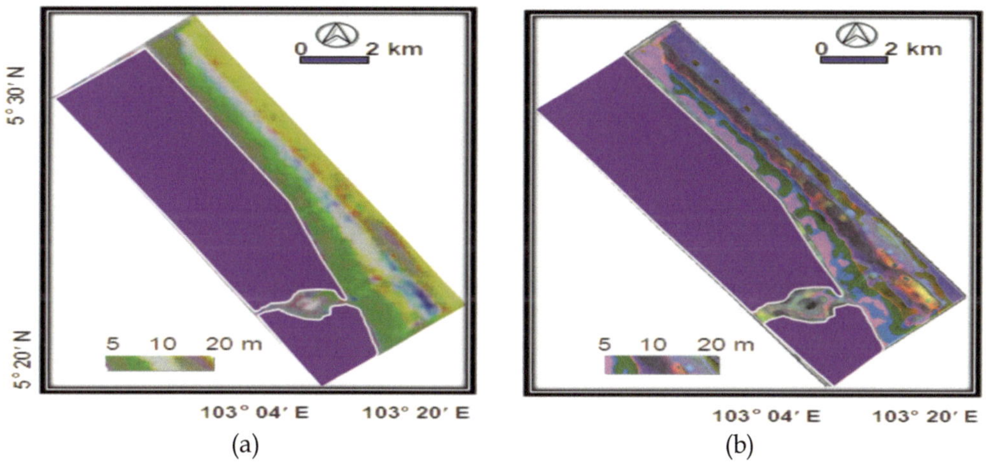

Fig. 4. Result of anisotropic diffusion algorithm for bathymetry signature from (a) C_{VV} and (b) L_{HH} bands.

Comparison between Figs. 5 and 6 showed that the L_{HH} band captured a stronger tidal current flow than the C_{VV} band. The maximum tidal current velocity simulated from the L_{HH} band is 1.6 m s^{-1} while the ones is simulated from C_{VV} band is 1.4 m s^{-1}. This is because different bands with differrent polarizations. The major axis of tidal current is towards the south and approximately moving parallel to shoreline (Figs. 5 and 6). In addition, it is obvious that both bands are imaging the major axis of tidal current in the range direction.

This is because December represents the northeast monsoon period as the coastal water currents in the South China Sea tend to move from the north towards the south (Marghany 1994). The travelling the of current is caused by the weak non-linearity due to the smaller value of *R/V*. The weak non-linearity was assisted by the contribution of the linear Volterra kernels of the range current. This means that the range current will be equal to zero when the Volterra kernels $H_{1y}(v_x,v_y)$ of the frequency domain has a zero for $\left|v_x \text{ and } v_y\right|$. However, the inversion of the linear kernel of the Volterra algorithm allowed us to map the current movements along the range direction. This result confirms the study of Inglada and Garello (1999). The results of the Volterra algorithm showed that there was an interaction between water flow from the mouth of the Kuala Terengganu River and the near South China Sea water flow which appeared to be close to the mouth of the Kuala Terengganu River (Marghany 2009 and Marghany et al., 2010).

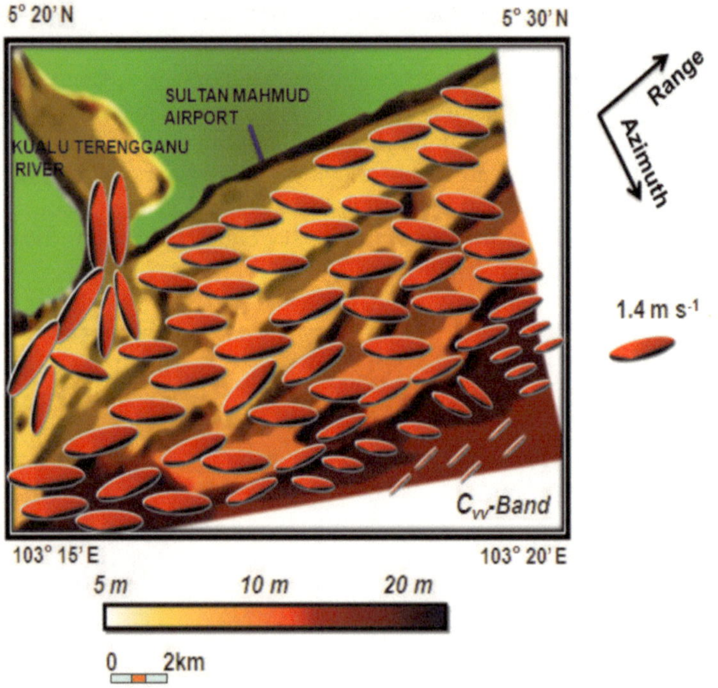

Fig. 5. Tidal Current Ellipses Simulated from C_{VV} band.

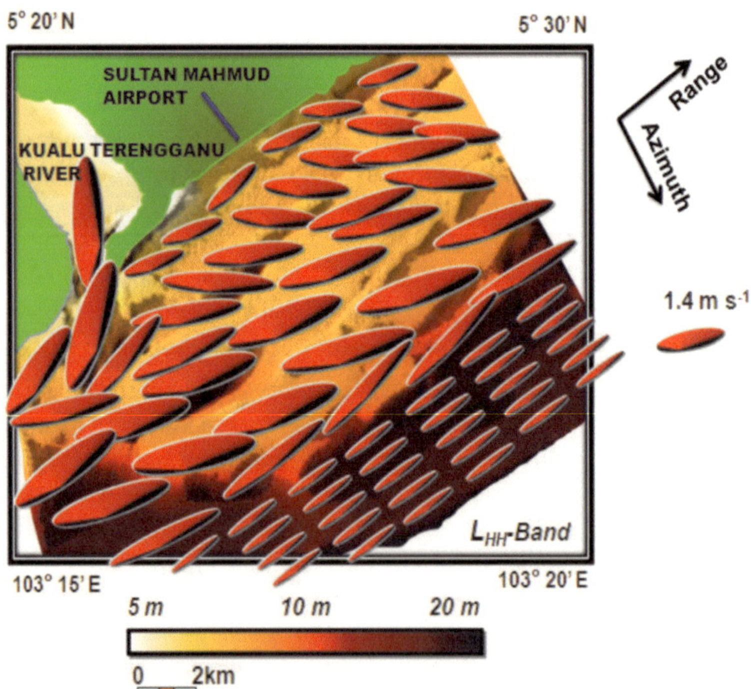

Fig. 6. Tidal Current Ellipses Simulated from L_{HH} band.

In addition, during the data acquisition time, the wind was blowing at about 8 m s^{-1} from northeast and swell system was propagated from the northeast. In this regard, the quality of bathymetry map simulated from C-band degrades because the image modulation become weaker relative to the speckle noise, and they are smeared out over a larger area due to the effect of long waves, which also add noise. As L-band data suffer less from these drawbacks, using of L-band provides more accurate results. In fact L-band has higher signal-to-noise ratio compared to C band. This confirms the study of Vogelzang (1997). Furthermore, the HH polarization has a larger tilt modulation compared to the VV polarization. Tilt modulation explains that the Bragg scattering is dependent on the local incident angle. The long wavelength of L-band HH polarization modulates this angle, hence modifying the Bragg resonance wave length. It might be due to the fact that the first - order Bragg Scattering gives good results for long radar wavelengths (L-band), but for shorter radar wavelength (C-band) the effects of waves longer than the Bragg waves must be taken into account (Shuchman et al., 1985 and Romeiser and Alpers, 1997). This could be due to strong current flow from the mouth river of the Kuala Terengganu. This study confirms the studies of Li et al., (2009) and Marghany et al., (2010).

Figure 7 shows the comparison between the 3-D bathymetry reconstruction from the topographic map , the L_{HH} band data, and the C_{VV} band data. 3-D topographic map was created using fuzzy concept by converting the 2-D topographic map into fuzzy interval number of [0,1]. It is obvious that the coastal water bathymetry along the Sultan Mahmud Airport has a gentle slope and the bathymetric contours are parallel to the shoreline. Close to the river mouth, the bathymetry at this location shows a sharp slope. The L_{HH} band captured a more real bathymetry pattern than the C_{VV} band. Further, Fig. 8 shows a clear discrimination between smooth and rough bathymetry where the symmetric three-dimensional structure of the bathymetry of a segment of a connecting depth. This can be noticed in areas A, B, C, D, E in real and L_{HH} band data compared to C_{VV} band. Smooth sub-surfaces appear in Figure 6 where the near-shore bathymetric contour of 5 m (area E) water depth runs nearly parallel in 3D-space to the coastline which is clear in Figure 6. Further, statistical analysis using regression model (Fig. 7) has confirmed that L_{HH} band tends to get closer to the true mean of real bathymetry map i.e. it is actual measured as more real, as compared to C_{VV} band data. A rough sub-surface structure appears in steep regions of 20 m water depth (areas of B, C, and D). This is due to the fact that the fuzzy B-splines considered as deterministic algorithms which are described here optimize a triangulation only locally between two different points (Anile et al. 1995). This corresponds to the feature of deterministic strategies of finding only sub-optimal solutions.

This result could be confirmed using linear regression model (Fig. 8). In this regard, Fig. 8a shows the regression relation between the observed bathymetry and the results obtained using the C_{VV} band TOPSAR data. Figure 8b shows a similar regression relationship for L_{HH} TOPSAR data. The scatter points in Fig. 8b are more close to the regression line than those in Fig. 8a. The bathymetry simulation from L_{HH} band with r^2 value of 0.95 and accuracy (root mean square) of ±0.023 m is more accurate than that obtained by using C_{VV} band with accuracy of (root mean square) ±0.03 m

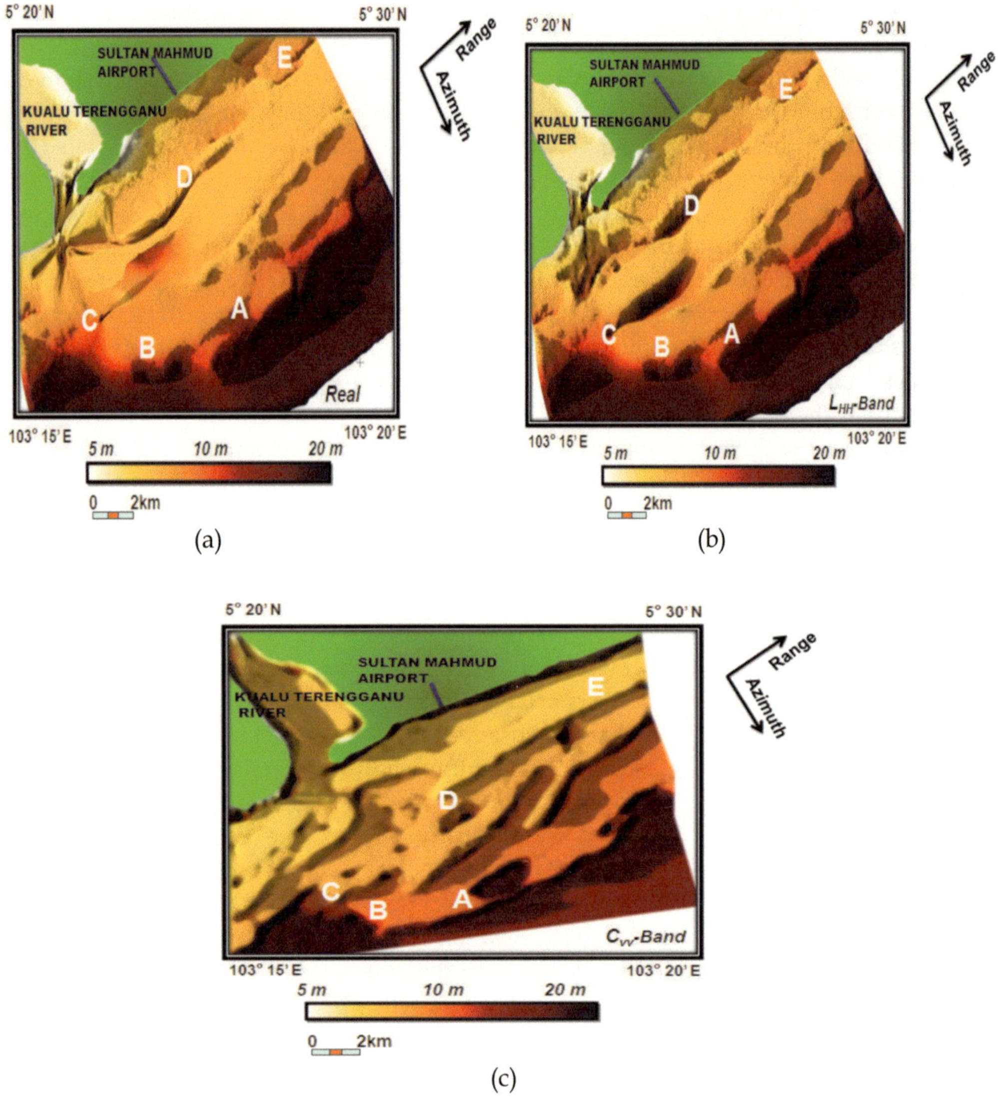

Fig. 7. Three-Dimensional Bathymetry Reconstructions from (a) Real Topography Map (b) L_{HH} Band and (c) C_{VV} Band.

It is clear that involving of fuzzy B-spline in 3-D bathymetric mapping has produced accurate bathymetry visualization. Therefore, the sharp visualization of 3-D bathymetry with the different TOPSAR polarised bands and real data due to the fact that each operation on a fuzzy number becomes a sequence of corresponding operations on the respective μ-levels, and the multiple occurrences of the same fuzzy parameters evaluated as a result of the function on fuzzy variables (Anile, 1997, Anile et al. 1997). It is very easy to distinguish between smooth and jagged bathymetry. Typically, in computer graphics, two objective quality definitions for fuzzy B-splines were used: triangle-based criteria and edge-based criteria. Triangle-based criteria follow the rule of maximization or minimization, respectively, of the angles of each triangle (Fuchs et al. 1997) which prefers short triangles with obtuse angles. This finding confirms the studies of Keppel (1975), Anile (1997), and Marghany et al., (2010).

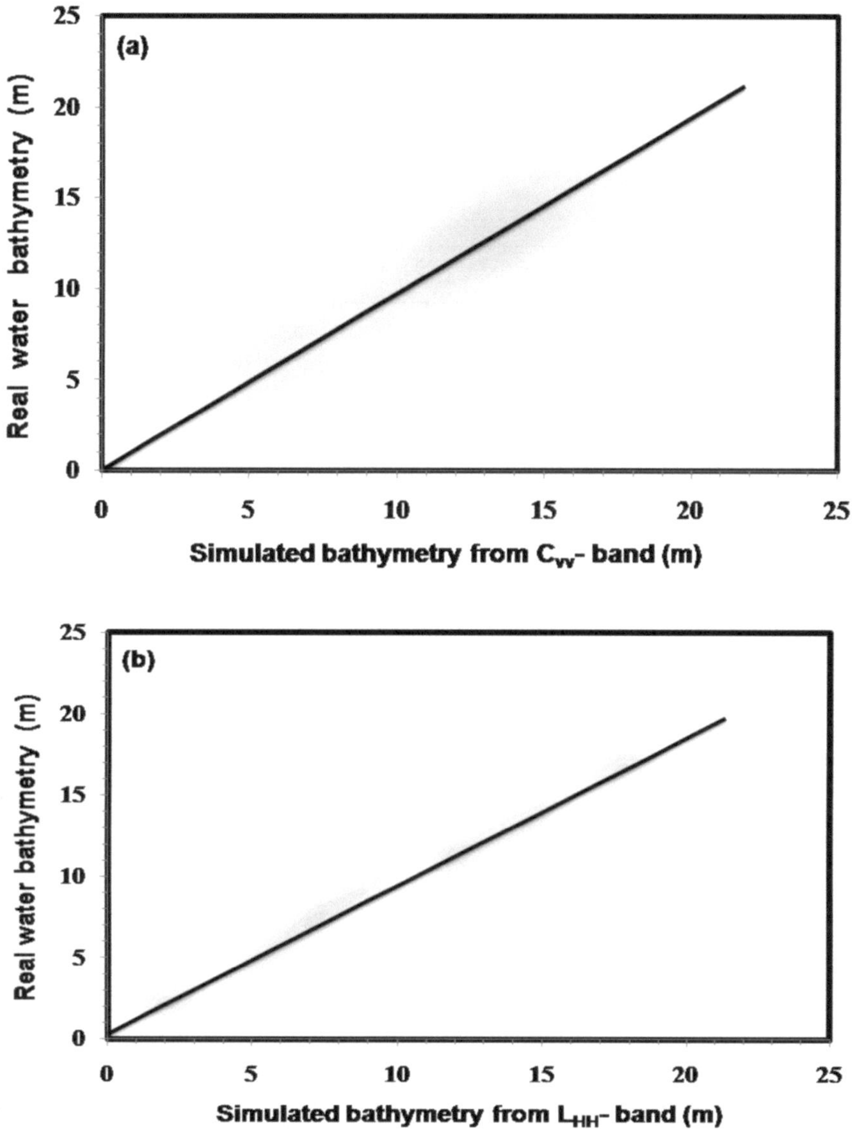

Fig. 8. Regression model between real water bathymetry from bathymetry Chart and (a) water Bathymetry from C_{VV} band (*r^2=0.85; y=0.95x+1.89 ;rms=±0.03 m*) and (b) L_{HH} band (*r^2=0.95; y=1.01 x+0.121;* rms=*±0.023 m*).

The three-dimensional bathymetry construction is not similar to the study of Inglada and Garello (1999), such that in the latter the bathymetry was constructed in the shallow sand waves (Garello 1999)due to the limitation of the inversion of the linear kernel of the Volterra algorithm. The integration of the inversion of the Volterra algorithm with fuzzy B-splines improved the three-dimensional bathymetry reconstruction pattern. The result obtained in this study disagrees with the previous study by Inglada and Garello (1999 and 2002) who implemented two-dimensional Volterra model to SAR data. 3-D object reconstruction is required to model variation of random points which are function of *x,y,* and *z* coordinates rather than using two coordinates i.e. (*x,y*). In addition, finite element model is required to discretize two-dimensional Volterra and continuity models in study of Inglada and Garello (1999 and 2002) to acquire depth variation in SAR image without uncertainty. Previous studies done by Alpers and Hennings (1984); Shuchman et al. (1985); Vogelzang (1997); Romeiser and Alpers (1997); Hesselmans et al., (2000); and Li et al., (2009) were able to model spatial variation of sand waves.

Splinter and Holman (2009) have developed algorithm that is based on the changing direction of refracting waves to determine underlying bathymetry gradients function of the irrotationality of wavenumber condition. In this context, Splinter and Holman (2009) claimed that depth dependences are explicitly introduced through the linear dispersion relationship. Further, they used spatial gradients of wave phase and integrated times methods between sample locations (a tomographic approach) to extract wave number and angle from images. They found that synthetic bathymetries of increasing complexity showed a mean bathymetry bias of 0.01 m and mean rms of 0.17 m. Nevertheless, refraction-based algorithm has limitations in which it can be applied only within 500 m from the shoreline. In this circumstance, bathymetry of complex rough sea surface interaction cannot be determined. This suggests that the refraction-based algorithm is best suited for shorter period swell conditions in intermediate water depths such as a semi-enclosed sea. Further, the refraction-based algorithm cannot be implemented in SAR data. In fact, the shortest wavelength less than 50 m cannot be estimated in SAR data due to the limitation of using two dimensional (2D) Fourier transform (Romeiser and Alpers, 1997).

In this study, fuzzy B-spline algorithm produced 3-D bathymetry reconstruction without existence of shallow sand waves. In fact fuzzy B-spline algorithm is able to keep track of uncertainty and provide tool for representing spatially clustered depth points. This advantage of fuzzy B-spline is not provided in Volterra algorithm and 1-D or 2-D continuity model.

6. Conclusions

Coastal bathymetry is tremendous information for coastal engineering, coastal navigation, economic activities, security and marine environmental protection. Single- or multi-beam ship-borne echo sounders are the conventional techniques used to map ocean bathymetry. Indeed, SAR data can reduce the root mean square error of bathymetry mapping from conventional methods by overall of 40 %. In this paper, we address the question despeckles' impact on the accuracy of depth determination in TOPSAR data without needing to include any sounding data values. This verified with airborne SAR data (namely the TOPSAR) using integration of the anisotropic diffusion algorithm ,the Volterra kernel and the fuzzy B-spline algorithm. Incidentally, the inverse of Volterra algorithm then performed to retrieve 2-D tidal current flows from C_{VV} and L_{HH} bands. Besides, the 2-D continuity equation then used to retrieve the water depth. To retrieve 3-

D bathymetry pattern, the fuzzy B-spline has performed to 2-D water depth information which estimated using 2-D continuity equation.

The study shows that anisotropic diffusion algorithm provides a clear bathymetry signature in L_{HH} band data compared to C_{VV} band data. Further, the maximum tidal current flow simulated from the C_{VV} band was 1.4 m s^{-1} while the ones was simulated from L_{HH} band was 1.6 m s^{-1}. This was assisted L_{HH} band to capture more accurately bathymetry features with r^2 value of 0.95, standard error mean of ±0.023 m. In comparison with SAR satellite data, L_{HH} band performs better because of TOPSAR data acquired with HH-,VV-, HV- and VH-polarized signals from 5 m x 5 m pixels. Further, TOPSAR data provides digital elevation model (DEM) with RMSE ± 1 m in the near range to greater than ±3 m in the far range. Conventional survey, however, has lower resolution than L_{HH} band. Indeed, the conventional survey cover swath width of 37.5 m to 400 m. Nevertheless, L_{HH} band has limitation to detect more than 20 m water depth.

It can be said that the L_{HH} band provides a better approximation to the real shallow bathymetry than does the C_{VV} band. In conclusions, the integration between anisotropic diffusion algorithm, the Volterra algorithm and the fuzzy B-splines could be an excellent tool for 3-D bathymetry determination from TOPSAR polarized data.

7. References

Alpers, W., and Hennings, I., (1984). A theory of the imaging mechanism of underwater bottom topography by real and synthetic aperture radar, *Journal of Geophysical Research*, 89, 10,529-10,546.

Anile, A. M, (1997). *Report on the activity of the fuzzy soft computing group*, Technical Report of the Dept. of Mathematics, University of Catania, March 1997, 10 pages.

Anile, AM, Deodato, S, Privitera, G, (1995) *Implementing fuzzy arithmetic*, Fuzzy Sets and Systems, 72,123-156.

Anile, A.M., Gallo, G., Perfilieva, I., (1997). *Determination of Membership Function for Cluster of Geographical data.* Genova, Italy: Institute for Applied Mathematics, National Research Council, University of Catania, Italy, October 1997, 25p., Technical Report No.26/97.

Forster, B.C., (1985). Mapping Potential of Future Spaceborne Remote Sensing System. Procs. of 27th Australia Survey Congress, Alice Springs, Institution of Surveyors, Australia, Australia, 109-117.

Fuchs, H. Z.M. Kedem, and Uselton, S.P., (1977). Optimal Surface Reconstruction from Planar Contours. *Communications of the ACM*, 20, 693-702.

Guenther, G.C., Cunningham, A.G., LaRocque, P. E., and Reid, D. J. (2000). Proceedings of EARSeL-SIG-Workshop LIDAR,Dresden/FRG,EARSeL , Strasbourg, France,June 16 - 17, 2000.

Hesselmans, G.H, G.J Wensink C.G. V. Koppen, C. Vernemmen and C.V Cauwenberghe (2000). Bathymetry assessment Demonstration off the Belgian Coast-Babel. The Hydrographical Journal . 96: pp.3-8.

Inglada, J. and Garello R.,(1999). Depth estimation and 3D topography reconstruction from SAR images showing underwater bottom topography signatures. In *Proceedings of*

Geoscience and Remote Sensing Symposium, 1999, *IGARSS'99, Hamburg, Germany, 28 June-2 July 1999, IEEE Geoscience and Remote Sensing Society, USA* . 2:pp. 956-958.

Inglada, J. and Garello, R., (2002). On rewriting the imaging mechanism of underwater bottom topography by synthetic aperture radar as a Volterra series expansion. IEEE Journal of Oceanic Engineering. 27, pp: 665-674.

Keppel, E. (1975). Approximation Complex Surfaces by Triangulations of Contour Lines. *IBM Journal of Research Development*, 19, pp: 2-11.

Lee, J. S., D. Schuler, T. L. Ainsworth, E. Krogager, D. Kasilingam, M.A. and Boerner, W.M., (2002). On the estimation of radar polarization orientation shifts induced by terrain slopes, IEEE Transactions on Geosciences and Remote Sensing, 40, pp: 30–41.

Li, X., L., Qing Xu, and Pichel, W. G., (2009). Sea surface manifestation of along-tidal-channel underwater ridges imaged by SAR. IEEE Transactions on Geosciences and Remote Sensing. 8, pp: 2467-2477.

Lyzenga, D.R., P. M. Norman, and Fred, J. T., (2006). Multispectral bathymetry using a simple physically based Algorithm. IEEE Transactions on Geosciences and Remote Sensing, 8, pp: 2251-2259.

(Malaysian Department of Survey and Mapping 1996). "Chendering". In tidal table. Pp. 241-264.

Marghany, M., (1994). Coastal Water Circulation off Kuala Terengganu, Malaysia". MSc. Thesis Universiti Pertanian Malaysia (now Universiti Putra Malaysia).

Marghany, M., (2005).Fuzzy B-spline and Volterra algorithms for modelling surface current and ocean bathymetry from polarised TOPSAR data. *Asian Journal of Information Technology*. 4, pp: 1-6.

Marghany M., and Hashim, M.,(2006). Three-dimensional reconstruction of bathymetry using C-band TOPSAR data. Photogrammetrie Fernerkundung Geoinformation. pp: 469-480.

Marghay, M., M., Hashim and Crackenal, A., (2007). 3D Bathymetry Reconstruction from AIRBORNE TOPSAR Polarized Data. In: Gervasi, O and Gavrilova, M (Eds.): Lecture Notes in Computer Science. Computational Science and Its Applications – ICCSA 2007, ICCSA 2007, LNCS 4705, Part I, Volume 4707/2007, Springer-Verlag Berlin Heidelberg, pp. 410–420, 2007.

Marghany M (2009). Volterra - Lax-wendroff algorithm for modelling sea surface flow pattern from Jason-1 satellite altimeter data. Lecture Notes in Computer Science (including subseries Lecture Notes in Artificial Intelligence and Lecture Notes in Bioinformatics) Volume 5730 LNCS, 2009, Pages 1-18.

Marghany, M. S., Mansor and Hashim, M., (2009a). Geologic mapping of United Arab Emirates using multispectral remotely sensed data. American J. of Engineering and Applied Sciences. 2, pp: 476-480.

Marghany,M., M. Hashim and Cracknell A (2009b). 3D Reconstruction of Coastal Bathymetry from AIRSAR/POLSAR data. Chinese Journal of Oceanology and Limnology.Vol. 27(1), pp.117-123.

Marghany, M. and M. Hashim (2010). Lineament mapping using multispectral remote sensing satellite data. International Journal of the Physical Sciences Vol. 5(10), pp. 1501-1507.

Marghany, M., M. Hashim and Cracknell A. (2010). 3-D visualizations of coastal bathymetry by utilization of airborne TOPSAR polarized data. *International Journal of Digital Earth,* 3(2):187 - 206.

McBean, E.A., and Rovers, F.A., (1998), Statistical procedures for analysis of environmental monitoring data and risk assessment. Prentice Hall PTR Environment and Engineering Series, Vol. 3. Upper Saddle River, New Jersey 07450. pp. 33-35.

Maeda, J., Iizawa, T., Tohru,I., and Suzuki,Y., (1997). Accurate segmentation of noisy Images Using Anisotropic Diffusion and linking of Boundary edge. IEEE TENCON-Speech and Image Technology for Computing and Telecommunications, Vol. 1,279-282.

Melba M., Kumar S., Richard M.R., Gibeaut J.C. and Amy N., (1999), Fusion of Airborne polarmetric and interferometric SAR for classification of coastal environments. *IEEE Transactions on Geosciences and Remote Sensing,* 37, pp: 1306-1315.

Mills, G. B., (2006) NOAA, Office of Coast Survey, Hydrographic Surveys Division, 1315 East-West Highway, Station 6859, Silver Spring, Maryland, USA 20910-3282. (Url: http://chartmaker.ncd.noaa.gov/hsd/ihr-s44.pdf, accessed December 2006).

Perona, P. and Jitendra Malik (1987). "Scale-space and edge detection using anisotropic diffusion". Proceedings of IEEE Computer Society Workshop on Computer Vision,. pp. 16–22.

Perona,P. and Jitendra Malik (1990). "Scale-space and edge detection using anisotropic diffusion". IEEE Transactions on Pattern Analysis and Machine Intelligence, 12 (7): 629–639.

Romeiser, R. and Alpers, W., (1997), An improved composite surface model for the radar backscattering cross section of the ocean surface, 2, Model response to surface roughness variations and the radar imaging of underwater bottom topography, *Journal of Geophysical Research,* 102, pp: 25,251-25,267.

Russo, F., (1998).Recent advances in fuzzy techniques for image enhancement. IEEE Transactions on Instrumentation and measurement. 47, pp: 1428-1434.

Rövid, A., Várkonyi, A.R. andVárlaki, P., (2004). 3D Model estimation from multiple images," IEEE International Conference on Fuzzy Systems, FUZZ-IEEE'2004, July 25-29, 2004, Budapest, Hungary, pp. 1661-1666.

Sapiro,G. (2001). Geometric partial differential equations and image analysis. Cambridge University Press. p. 223.

Shuchman, R.A., Lyzenga, D.R. and Meadows, G.A. (1985). Synthetic aperture radar imaging of ocean-bottom topography via tidal-current interactions: theory and observations, *International Journal of Remote Sensing,* 6, 1179-1200.

Splinter, K.D., and Holman R.A., (2009). Bathymetry Estimation from Single-Frame images of nearshore waves. IEEE Transactions on Geosciences and Remote Sensing, 47, pp: 3151–3160.

Yu, Y., and T. A., Scott (2002), Speckle reducing anisotropic diffusion. *IEEE Transactions on Geosciences and Remote Sensing,* 11, 1260-1270.

Vogelzang, J. (1997)., Mapping submarine sand waves with multiband imaging radar, 1, Model development and sensitivity analysis, *Journal of Geophysical Research,* 102, 1163-118.1.

Vogelzang, J., Wensink, G.J., Calkoen, C.J. and van der Kooij, M.W.A. (1997). Mapping submarine sand waves with multiband imaging radar, 2, Experimental results and model comparison, *Journal of Geophysical Research*, 102, 1183-1192.

Vogelzang, J., Wensink, G.J., de Loor, G.P., Peters, H.C. and Pouwels, H., (1992). Sea bottom topography with X band SLAR: the relation between radar imagery and bathymetry, *International Journal of Remote Sensing*, 13, pp: 1943-1958.

Wensink, H. and Campbell, G., (1997). Bathymetric map production using the ERS SAR. *Backscatter*, 8, pp: 17-22.

Part 2

Applications of Bathymetry

4

The Geomorphology and Nature of Seabed Seepage Processes

Martin Hovland
University of Bergen/Statoil ASA
Norway

1. Introduction

The seafloor is mostly flat and smooth, reflecting sedimentation, the dominating seabed formation mode. In this communication, I shall focus on how vertical fluid flow, also called "seepage" may alter and imprint the seafloor geomorphology. Seafloor seeps occur globally, wherever a fluid (gas or liquid) resides in the sub-strata below the seafloor and finds a way up. The fluids may come up from depths of several kilometres below surface or from shallower depths. Because they have lower specific gravity than their surroundings, they tend to move through the pore-spaces of sediments or fissure-networks of solid rocks, and upwards towards the surface. At locations where they break through the seafloor, depressions and sometimes mounds may form (Fig. 1).

Fig. 1. Seepage of fluids (liquids and gas) though the seafloor may occur through positive seafloor relief (left arrow) or it may induce negative relief features in the seafloor (right arrow).

2. The "Second Surface" with seep and vent manifestations

The seafloor is the "Second Surface" of our Planet, indicating that it is hidden in many ways. Totally, it covers an area which is about 3 times larger than the land-surface. Paradoxically, there exists more visual documentation of the total surfaces of both Moon and Mars, than of the immensely more important combined Land- and Second Surfaces of Earth. From sediment sampling, fishing (trawling), scientific dredging (dredge sampling), and drilling, in all oceans, over time, it is currently known that the Second Surface mostly consists of mud (clay), sand, rock, and in some areas metals and salts. But, because it is flooded by water, it is both pressurized, and buoyed at the same time, and behaves accordingly. On average, the Second Surface has a much thinner crust (up to 15km thick) than the on-shore continental crust (up to 200 km thick) and is, therefore, more likely to be exposed to high heat-flow from the Earth's interior. Thus, along tectonic plate boundaries (mid-ocean spreading zones, transform faults, and subduction zones) the high heat-flow induces venting of warm and hot fluids in so-called 'hydrothermal vent systems', which will only be treated briefly in this chapter.

Generally, the ocean floor is covered in thick sediments that deposit by gravitation, with particles that sink through the water column and accumulate in thick layers on the Second Surface. The fluids, including petroleum gas and liquids (hydrocarbons) trapped underneath such sediments are lighter than the solids and, therefore, move upwards to surface at discrete locations due to buoyancy. This process is also called "migration" and where the flow penetrates the Second Surface from below it is called "fluid flow" (e.g. Judd and Hovland, 2007). The discrete locations where the fluids occur at the surface are called 'cold seep' locations (seep 'manifestations'). Depending on the geological setting, the distance between each cold seep location on the seafloor varies considerably, from kilometres, to only several metres. However, cold seeps are important for life within, on, and above the Second Surface because they represent transport pathways for dissolved chemical constituents including nutrients and can, therefore, sustain unique oasis-type ecosystems at the seafloor (e.g. Hovland, 1984; Hovland and Judd, 1988). Fluids expelled through seeps contain re-mineralized nutrients (silica, phosphate, ammonia, and alkalinity) and hydrogen sulphide, as well as dissolved and free methane from microbial degradation of sedimentary organic matter. Because methane gas molecules (CH_4) have the highest relative hydrogen content (4 hydrogen atoms to 1 carbon atom) of any organic compound, they represent a valuable energy source to certain primary producers: archaea and bacteria, i.e., the methanotrophs and the methane oxidizers. Apart from near cold seep locations, seawater has generally very low concentrations of methane and other light hydrocarbons, such as ethane (C_2H_6), propane (C_3H_8), butane (C_4H_{10}), and pentane (C_5H_{12}). Perhaps the single most important reaction associated with cold seeps, is the anoxic oxidation of methane (AOM) by archaea and sulphate reducing bacteria (SRB), with secondary reactions involving the precipitation of carbonate ($CaCO_3$), in the form of inorganic aragonite and calcite (Hovland et al., 1985). Reeburg (2007) pointed out the complexity of the biogeochemistry of oceanic methane circulation and stated: "A geochemical budget is a flux balance that provides a useful means of partitioning and estimating the magnitudes of sources and sinks. Budgets are very useful in exposing our ignorance, but they have no predictive power." Even though there is still a high uncertainty about the impact of methane seepage from the seafloor, this is certainly one of the emerging areas for scientific studies.

In contrast to the petroleum-related natural seeps, the fluids that seep up from volcanic activity in the deep-ocean are often a mixture of carbon dioxide (CO_2), methane (CH_4), and hydrogen (H_2). Organisms that feed on nutrients brought up in such deep-water seeps can build large structures on the seafloor made out of carbonate rock ($CaCO_3$). These structures are called "reefs" and mounds. Figure 2 provides a remarkable illustration of some ancient seep-related carbonate mounds that grew on a now uplifted seafloor located in the Sahara desert, of all places. These fossil carbonate formations occur in the eastern Anti-Atlas mountains, southern Morocco (Belka, 1998).

Fig. 2. Left: Some ancient, spectacular fossil coral reefs as found in the Sahara desert today, after they have been buried by sand for many millions of years (Wendt et al., 1997). Right: A rendering of how they might have looked like, when alive, some 380 million years ago.

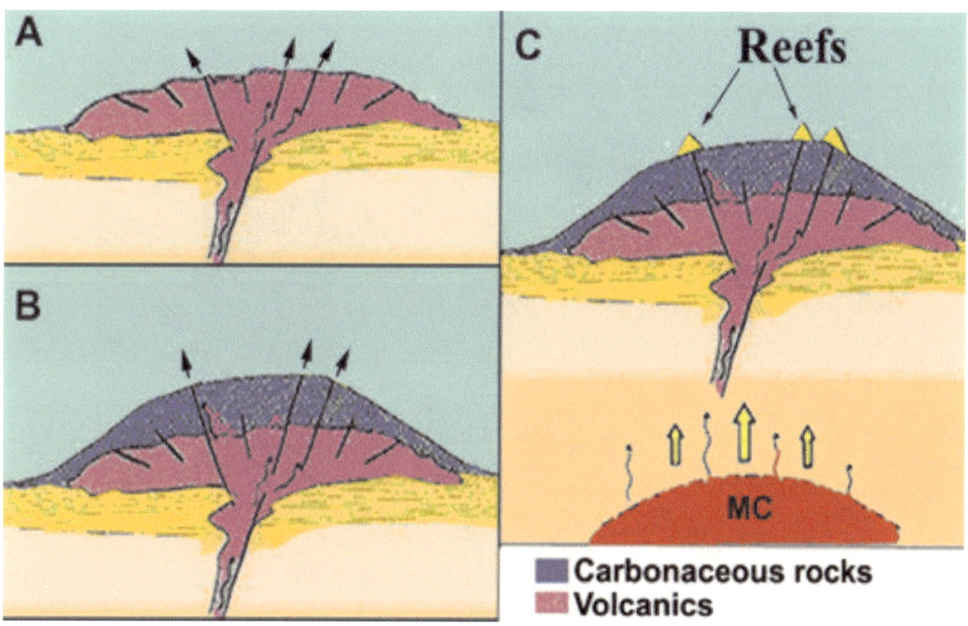

Fig. 3. Conceptual cartoon (based on Belka, 1998) indicating in chronological order (A, B, C) how volcanic outflow onto the seafloor may have stimulated the growth of the carbonate mounds (yellow cones, denoted "Reefs") of Sahara. "MC"= magma chamber.

They are of Early Devonian age and are possible to see only because they occur in a desert landscape, without vegetation. In all, there are 54 such structures within the Hamar Lagdhad area of Sahara (Hovland, 2008). The fossils within these ancient Saharan mounds include trilobites and crinoids. There are also corals, brachiopods, ostracods, gastropods, and pelecypods. Even though there are markings from worms and sponges, there are no signs of algae. This indicates that the structures have been growing at depths greater than 100 m (Belka, 1998). A special character of these mounds, are the numerous "neptunian dikes", fracture systems, which have been filled with other sediments and minerals. It is suspected that these dikes represent spring conduits (seeps) during the formation of the mounds, and that the seeping fluids were of a hydrothermal nature (Fig. 3), which must have stimulated the organic growth in the first place.

3. Seeps in the North Sea

Within the central and northern North Sea, there are three, fairly well-studied methane seep locations: the Tommeliten seep area (56°29.90′ N, 2°59.80′E) (Hovland and Sommerville, 1985; Hovland and Judd, 1988; Niemann et al., 2005; Wegener et al., 2008; Schneider von Deimling, 2010), the Scanner pockmark seeps (58°28.5′ N, 0°96.7′E) (Hovland and Sommerville, 1985; Hovland and Thomsen, 1989; Dando and Hovland, 1992), and the Gullfaks seeps (61°10.1′ N, 2°15.8′E) (Hovland, 2007; Wegener et al., 2008). Although each of them are located in different geological settings, they have one main aspect in common: - they occur as continuous macro-methane seeps (Fig. 4). Except for the Scanner seeps, the

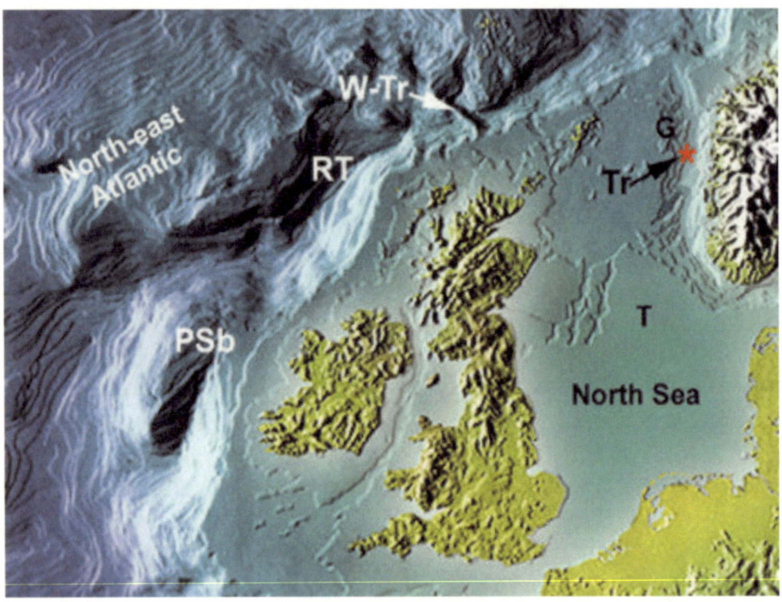

Fig. 4. The general bathymetry of the North East Atlantic Ocean including the North Sea. Seep locations in the North Sea are shown on the map as: T=Tommeliten, G=Gullfaks, and Tr=Troll. The Scanner seeps are located just north of T. The other features indicated, PSb=Porcupine Seabight, RT=Rockall Trough, and W-TR=Wyville-Thomson Ridge (based on Hovland, 2008). Because the mean sea-level during the Last Glacial Maximum (LGM) was about 120 m lower than at Present (Jelgersma, 1979) and because the three mentioned seep locations occur at three different water depths, the seafloor sediments at these locations geologically span the transition zones from *terrestrial* (Tommeliten, at ~75 m water depth, wd), through *intertidal* (Gullfaks, 130 m wd) to *submerged marine* (Scanner, 160 m wd) , and Troll, ~300 m wd zones.

two others have, over the last 20 years, been studied by multidisciplinary (including microbial) scientific surveys, with some interesting and also generalizable results.

3.1 Seeps at Tommeliten

A research cruise was mobilized by Statoil, in 1983, to perform a first-hand assessment of the seep features at Tommeliten. The vessel 'Skandi Ocean' was equipped with towed side scan sonar, combined with sub-bottom (high-resolution) profiler, an ROV (remotely operated vehicle) capable of acquiring gas samples, and a gravity coring system (Hovland and Judd, 1988; Judd and Hovland, 2007). The seafloor at Tommeliten has a mean depth of 75 m, and is generally flat and even. It consists of a ca. 5 cm layer of fine to medium quartzite sand, overlying stiff clay and marl layers. As documented by the several km^2 area surveyed with the towed system, a much smaller, 0.25 km^2, area was surveyed visually, with the ROV (Fig. 5). Here, a total of 22 actively bubbling seeps were observed. However, the whole seep area which is about 100 m in diameter is estimated to contain 120 individual bubbling seeps. The total methane output from this seepage field, was estimated to be 24 m^3 of methane per day, at 75 m water depth (Hovland and Sommerville, 1985; Hovland et al., 1993). These were all concentrated within an area of about 0.06 km^2 (Wegener et al., 2008).

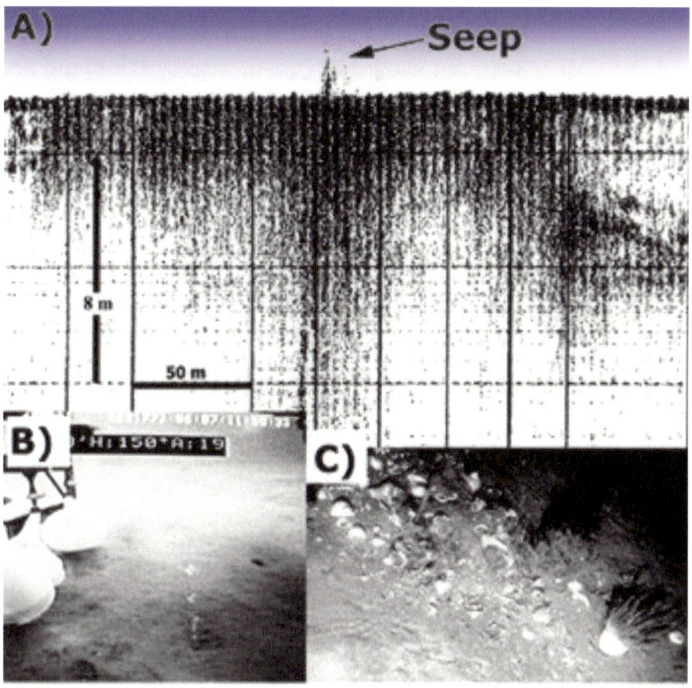

Fig. 5. A) One of the many bubbling seeps at Tommeliten, as recorded with the high-resolution sub-bottom profiler (SBP, 5 kHz). Note the gas-charged sub-surface sediments, which provide a strong backscatter (dark) acoustic signature. B) A bubbling seep at Tommeliten, seen in ambient light conditions, June, 1983. The size of the white funnels are about 20 cm high (Hovland and Judd, 1988). C) A small 'bioherm' located inside an 'eyed pockmark' near the seep depicted on A and B. The sea-anemone on the lower right is about 10 cm high (from Hovland and Thomsen, 1989).

The ebullition at Tommeliten is easily detectable with acoustic systems. However, these mid-water hydroacoustic "flares" (named thus because they are highly visible on echosounder recordings) are not only caused by various sized bubbles, but possibly also by water density contrasts caused by high concentrations of dissolved methane in the water. Thus, Niemann et al. (2005), reported up to 2 orders of magnitude higher concentrations (500 nM) of methane within the acoustic plumes (flares), compared to the background methane concentration (5 nM). Other possibilities for the formation of large, acoustic flares will be discussed later in this chapter. In general, a consequence of the hydrogen sulphide transported with escaping gas and interstitial water to the seafloor surface from deeper strata is the formation of sulphur oxidizing bacteria, notably the filamentous *Beggiatoa*, *Thiothrix*, and *Thioploca* spp (Dando and Hovland, 1992). They normally occur over patches of the seafloor where the sub-surface sediments are charged with reduced gases. They also occur close to gas outlets and on the underside of rocks (often carbonates) brushed by venting gas bubbles (Brooks et al., 1979; Hovland and Thomsen, 1997). These bacteria utilize chemical energy from sulphide oxidation to fix carbon dioxide into organic matter (Nelson et al., 1995). The grazing of macrofauna on bacterial mats has also been observed (Stein, 1984; Hovland, 2007).

3.2 The Scanner seeps

The actively seeping Scanner pockmark in the Witch Grounds, near the Forties field in the UK sector of the North Sea was found during a drillsite survey (Hovland and Sommerville, 1985). Subsequent mapping of the area revealed that the 900 m long, 450 m wide, and 22 m deep pockmark has several other large active pockmarks in its neighbourhood. The seeps were first noticed as acoustic flares on hull-mounted echosounder and towed side scan sonar data. During an ROV-based survey conducted by Statoil, in 1985, they were acoustically detected with the vessel 'Lador' and visually localized with the ROV 'Solo'. However, compared to the acoustic flares, the bubble streams issuing from the Scanner pockmark were disappointingly small and feeble (Fig. 6). Only three bubble streams were found inside the pockmark and one of them, located adjacent to a protruding MDAC (methane-derived authigenic carbonate) block was sampled (Fig. 6). The maximum gas production (by bubble streams) was estimated to be 1 m^3 per day from the entire pockmark (Hovland and Sommerville, 1985).

During one ROV survey line across the active Scanner pockmark, in 1985, the ROV ran at a constant depth of about 130 m across the pockmark (Fig. 7). During this run, the ROV-mounted side scan sonar recorded some diffuse 'noise' on both sides of the vehicle. It looked like small parcels of water with contrasting density or acoustic reflectivity (caused by change in impedance). The survey vessel 'Lador', followed the 'Solo' and had the hull-mounted 38 kHz echosounder running. The 'Lador' echosounder recording shows the ROV beneath the vessel as a horizontal intermittent line. But a strong impressively large acoustic flare was centred over the pockmark (Fig. 7). During the horizontal survey transect through the water above the pockmark, no bubbles were seen on the ROV-acquired video from the water column in front of the ROV. Later, Wegener et al. (2008) observed a relatively large acoustic flare over the Scanner pockmark, which reached to about 80 m below the sea surface. The Scanner pockmark consists of a dense series or cluster of unit pockmarks (Hovland et al., 2010; Judd and Hovland, 2007), which becomes evident when the ROV moves from the outside over the outer rim of the pockmark. The landscape is undulating as the ROV descends down the gentle slope to the pockmark bottom, which lies 22 m below the surrounding seafloor. Because the seafloor consists of soft clay (mud), the ROV operation calls for careful navigation. Too much use of ROV-thruster energy renders the seawater murky and reduces visibility to less than 50 cm.

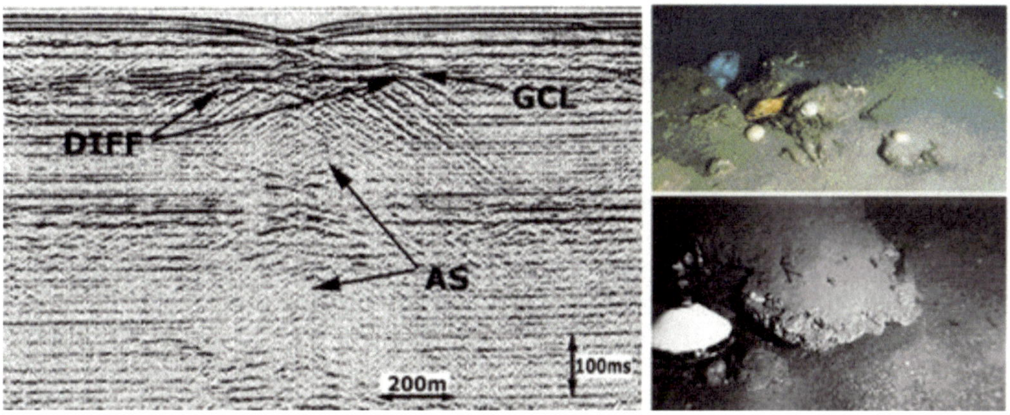

Fig. 6. The Scanner seep location is within a 22 m deep pockmark crater in the Witch Grounds of the North Sea, not far from the Forties field. The left image is a reflection seismic record showing reservoired gas (GCL) residing immediately below the pockmark depression. "Diff"=acoustic diffractions, "AS"= acoustic shadow zones. To the right are two images of the seafloor environment at the bottom of the pockmark. The lower one shows a white funnel (20 cm high), mounted on the ROV for gas sampling. Furthermore, there is a large MDAC rock adjacent to the bubbling gas stream being sampled (not visible). In the upper right photo, the red fish is about 20 cm long

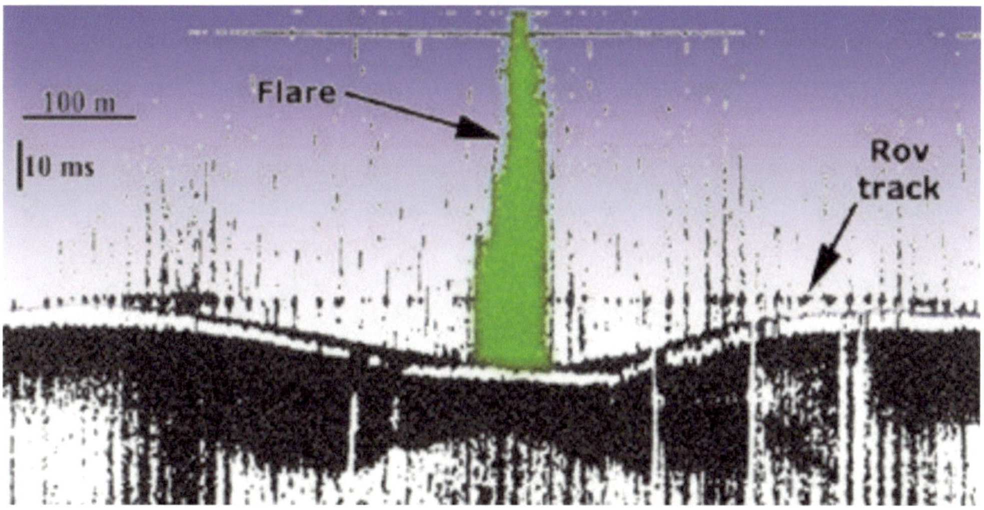

Fig. 7. A unique single beam echosounder record acquired over the Scanner pockmark by the vessel 'Lador' as the ROV 'Solo' surveyed the pockmark at a constant depth (~130 m) (see text for further details).

To explain the apparent mismatch between the feeble gas seepage observed and sampled inside the pockmark compared to the large hydroacoustic flare, it is suggested the flare is not only caused by rising bubbles, but may also be caused by high concentrations of methane and/or, perhaps, hydrogen (H_2). There is also another possibility: - that gas bubble clouds rising through the water column create a weak sound, i.e., noise that could be picked up as flares or weak reflections by echosounder and side scan sonar transducers. Either of these suggestions may be likely, but only careful acoustic and chemical studies will be able to determine how valid they are.

3.3 Pockmarks at Troll

The large Troll gas field is located at 310 m water depth inside the broad (>100 km wide) 'Norwegian Trough', which runs parallel with the southern and south-western coast of Norway (Fig. 4). There are numerous pockmarks in this area, which are up to 100 m in diameter and 8 m deep (Tjelta et al., 2007; Judd and Hovland, 2007). Despite the high density of up to 20 pockmark craters per km^2, at Troll, there are no known macro-seeps, detectable as hydroacoustic flares in the water column (Fig. 8). This probably means that most of the gas released in the area, is done episodically through the pockmarks. However, the periodicity of such release is still unknown (days, months, or years?).

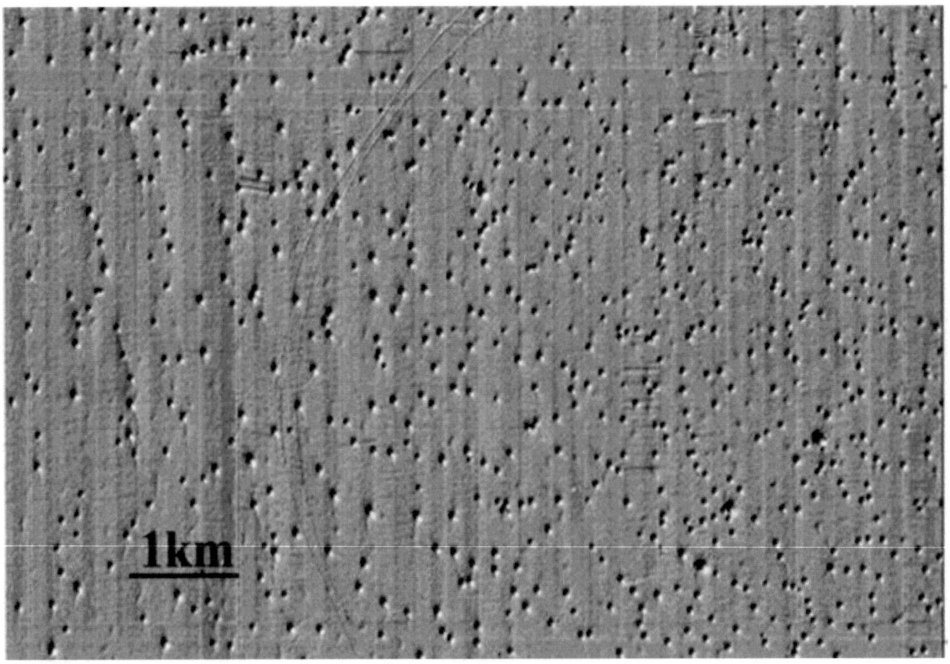

Fig. 8. Pockmarks mapped with hull-mounted multi-beam echo-sounder (MBE) near the Troll field, Northern North Sea. The resulting digital terrain model (DTM) is here presented as a shaded relief image with artificial lighting from the NW (north is up).

Statoil investigated some of the largest pockmarks over parts of the Troll field to find out more about the rate of natural sub-seafloor hydraulic activity (Forsberg et al., 2007) (Fig. 9). It is inferred that the eight 'satellite' pockmarks surrounding the parent-pockmark occur as a consequence of self-sealing of the parent, by the formation of MDAC across its bottom (see Fig. 9) (Hovland, 2002).

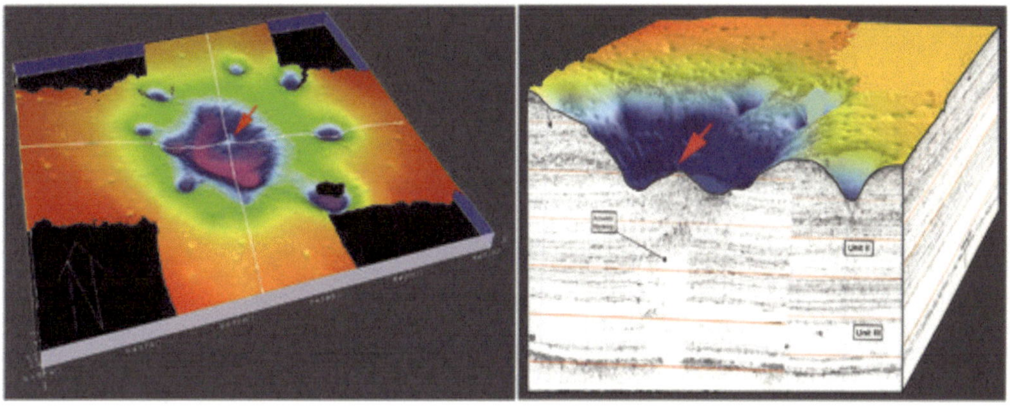

Fig. 9. The left image is a perspective view of a portion of the seafloor at the Troll field mapped at high resolution with ROV. It shows one central (old) 'parent pockmark' in the middle and eight of the other smaller 'satellite pockmarks' surrounding it. Right: Another view of the same features with enhanced vertical scale combined with high resolution SBP-data across the pockmarks. This drawing is made as a composite image based on real sub bottom profiler and bathymetric data from the actual location. The sides of this diagram are about 800 m, by 600 m, by 30 m. The red arrow points at the features seen in Fig. 10.

ROV-mounted sub bottom profiling (SBP) was also performed across this parent-pockmark. A vertical zone of disturbance and anomalous reflections beneath the centre of the pockmark were detected (Hovland et al., 2010). This zone is probably caused partly by the occurrence of MDAC and also a presence of small amounts of free gas, suspected to represent a cylindrical 'chimney' below the pockmark. On visual inspection with an ROV they came across some large colonies of soft coral (*Paragorgia* sp.) inside the largest pockmark inspected (Tjelta et al., 2007). To our knowledge, this is the first time large corals have been documented in the Norwegian Trough. The find came as a surprise as conditions are regarded to be far from perfect for such filter-feeding organisms. The two large *Paragorgia arborea* (one white and one red) individuals are perched inside the 8 m deep pockmark, which has a 1 m high conical methane-derived carbonate rock protruding up from its centre (Fig. 10). The corals are firmly based on this 'natural concrete' substratum. Clusters of up to 30 *Acesta excavata* bivalves are also affixed to the same structure. Because these animals live at a depth of up to 6 m below the general seafloor it is likely that they must tolerate frequent periods of heavy silting and sedimentation. These organisms undoubtedly occur here as a result of seepage-induced nutrient enrichment inside the pockmarks (Hovland et al., 1985).

Strings of small pockmarks were already identified on side scan sonar records in the middle of the Norwegian Trough during the early seafloor mapping activities conducted there (van Weering et al., 1973; Hovland, 1981, 1982; Hovland and Judd, 1988; Judd and Hovland, 2007). However, from the high-resolution mapping with ROV-mounted MBEs and side scan sonars conducted in 2005 (Fig. 11), the 'habitat' of unit-pockmarks in the near-Troll area has been documented. The most remarkable occurrences are those associated with clusters of normal-pockmarks, where one large pockmark occurs in association with several 'parasite' or satellite-pockmarks, as shown in Fig. 9 (Forsberg et al., 2007; Webb et al., 2009).

Fig. 10. Left: Large carbonate rock, a methane derived authigenic carbonate (MDAC). It serves as foundation for the *Paragorgians* and *Acesta* organisms seen in the right image. Notice the dense cluster of about 10 cm long *Acesta excavata* bivalves attached to the stems of the large gum corals and to the underlying MDAC-rock (From Hovland et al., 2010).

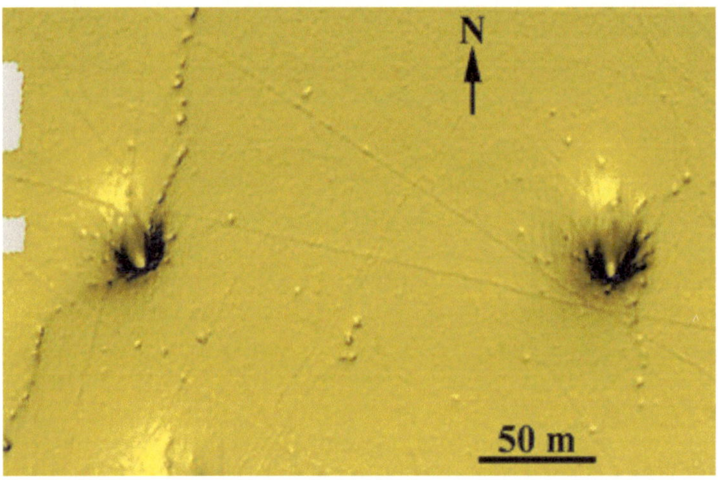

Fig. 11. High-resolution MBE data from the Troll-area (ROV-mounted MBE). This shaded relief DTM is presented in a light tan colour to enhance topography. Strings of pockmarks and trawl-marks are seen in this shaded relief image. The small (<5 m diameter) pockmarks making the strings are called "Unit pockmarks", and were defined by Hovland et al. (1984).

3.4 What are pockmarks?

In the foregoing, there has been a lot of mention about pockmarks, the mysterious craters in the seafloor. The strange fact is that they are about as mysterious today as they were when Lew King and Brian MacLean, back in the late 1960's discovered them (King and MacLean, 1970). The main reason for them not being understood yet, is that there are no such features on the terrestrial surface of our planet, only on the water covered Second Surface, which we have great difficulties in both imaging and understanding. Figures 8 and 12, however, give a good idea of how common they are on the seafloor in some areas, especially where there are hydrocarbons (gas and oil). One thing is certain, however, they have something to do with the hydraulics of the soft seafloor, which is both pressurized, and buoyed at the same time. One other important aspect is also that gases have something to do with their formation, as recently concluded by Cathles et al. (2010) and Hovland et al. (2010):

- "The local seafloor is either characterized as 'hydraulically active' or 'hydraulically passive', dependent on the occurrence of pockmarks and other fluid flow features (active) or the absence of such features (passive).
- The type of surface fluid flow manifestations determines the type and vigour of activity, i.e, cyclic/periodic, high, or low hydraulic activity.
- Whereas Unit-pockmarks most likely represent cyclic pore-water seepage, normal-pockmarks represent periodic or intermittent gas bursts (eruptions) with intervening periods of slow, diffusive, and cyclic pore-water seepage.
- The driving force behind seafloor hydraulic activity is reservoired, buried gas pockets.

In practical terms, this means that when sampling for seeping fluids, we recommend 'seep-hunters' to target the Unit-pockmarks. The higher investment needed to ensure detection and mapping of the small Unit-pockmarks, may be balanced by a higher success rate in sampling dissolved gases in the seeping pore-waters resulting from the active pumping by the trapped under-ground gas." (Hovland et al., 2010)

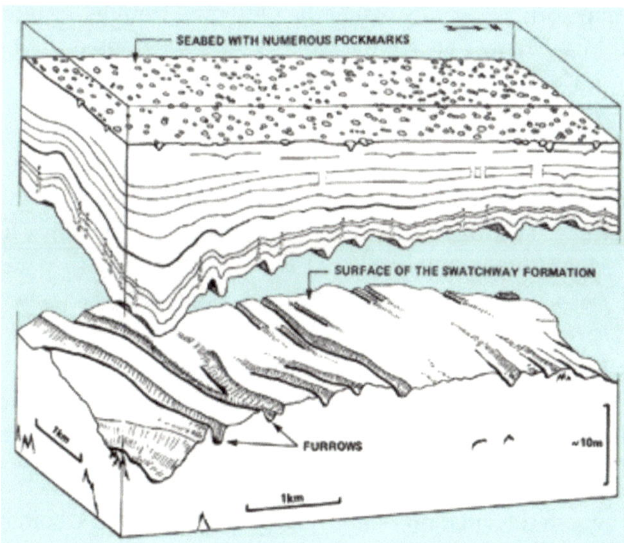

Fig. 12. Pockmarks in the Forties area, central North Sea (Based on Hovland and Judd, 1988).

4. Seeps off Mid-Norway

Part of the north-eastern Atlantic Ocean located to the west of Mid- and Northern-Norway is called the "Norwegian Sea". It stretches from the North Sea, north of the 62nd parallel to the islands of Spitzbergen. Over large areas, the seafloor off Mid-Norway and also parts of the Barents Sea, is still heavily iceberg plough-marked (Fig. 13). Thus, the geomorphology is very different from the seafloor over the majority of the North Sea and there is hardly any soft, layered sediments that have been laid down over the iceberg-scoured clay-rich sediments since the last glacial maximum (LGM) period about 20 thousand years ago (Judd and Hovland, 2007; Hovland, 2008). However, despite the dominating seabed topography being the kilometre long, up to 50 m wide and over 5 m deep linear troughs, there are more recent bedforms occurring, such as pockmarks, linear ridges and deep-water coral reefs (DWCRs), which have been partly or fully induced by fluid flow (seepage).

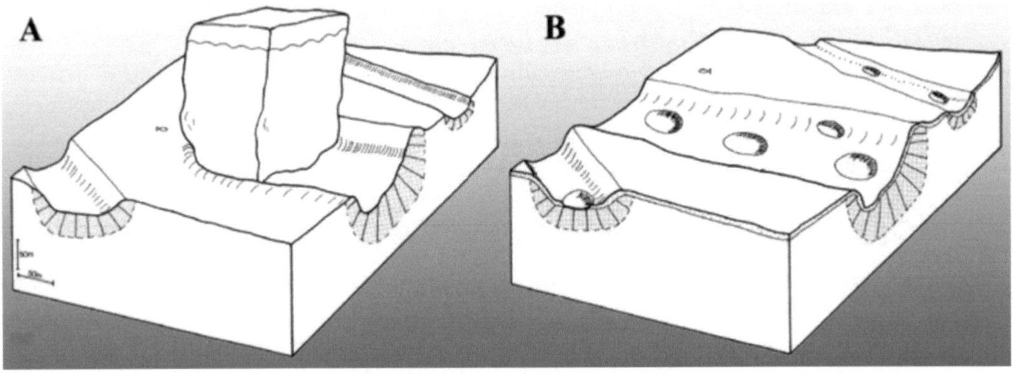

Fig. 13. A sketch showing how iceberg plough-marks were formed by grounded icebergs moving across the seafloor (A), during the melt-off period, after the LGM. Note that more recent pockmarks may form especially inside the ploughed troughs, as often seen on the present seafloor off Mid-Norway and in the Barents Sea (B) (adapted from Hovland and Judd, 1988).

4.1 Deforming gas-charged clay at Onyx?

Even though linear trends produced by drifting icebergs are very common offshore Mid-Norway, another, strange seafloor topography was found over the Onyx field, in concession block 6407/4, off Mid-Norway (Hovland, 2008). The more-or-less parallel ridges, there, resemble inverted iceberg plough-marks. These are, however, fundamentally different, as they are rounded upwards (Fig. 14 and 15). Furthermore, there are numerous seeps (detected as hydroacoustic flares) issuing from these ridges. The seafloor was only briefly investigated and no visual evidence of seepage was found on the seafloor. Only the occurrence of colonies of deep-water corals and an apparent bacterial mat (Fig. 16) were found to support the acoustic flare interpretations (Hovland, 1990a; 2008).

According to the suggested formation model, gas permeates into the soft, relatively stiff clay from below (shown as small dots in Fig. 14 , 1-3). The gas is expected to occur within the clay as finely disseminated minute bubbles, so that the bulk density of the gas-laden clay is less than that of the surrounding, gas-free clay. As more and more gas flows up from weakness zones in the underground, the upper surface layer (perhaps 20 m thick) starts to deform because of the bulk density (buoyancy) contrasts. As the deformation reaches a critical point, the clay becomes permeable and gas flows out to the water column, causing the observed hydro-acoustic flares. Gas will now follow these established conduits and the deformation ceases (Hovland, 1990a). A similar model was also used to explain the formation of linear, diapiric mud formations in the Adriatic Sea (Hovland and Curzi, 1989).

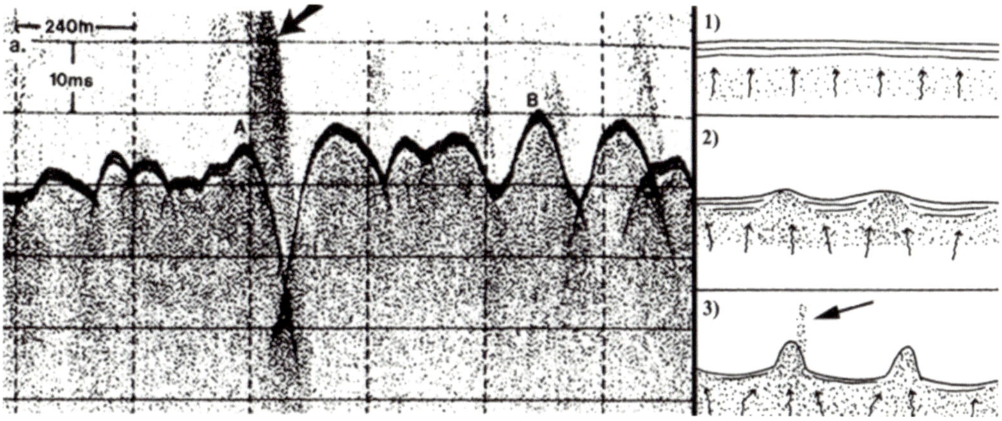

Fig. 14. Seepage and evidence of deforming clay at the Onyx field, off Mid-Norway (Based on Hovland 1990a and 2008). The sketches to the right, numbered 1 – 3 indicates a novel development model suggested by Hovland (1990a). See text for further description.

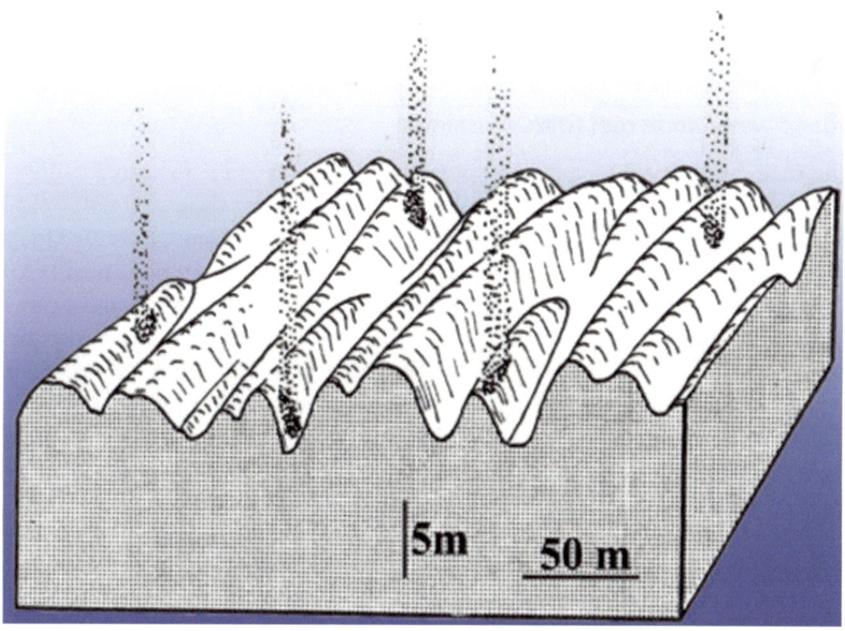

Fig. 15. A sketch showing how gas is suspected to flow out along linear ridges on the seafloor at the Onyx field off Mid-Norway. See text for further description and discussion (based on Hovland, 2008).

Figure 16 shows a coral accumulation at Onyx, near one of the seep locations. The field of view is about 1.5 m across at the base. Note the minute bacterial mat arrowed, which indicates the occurrence of reduced fluids in the sub-stratum. The organisms seen here are *Paragorgia arborea* (pink large coral, about 1 m high), some sponges, and a small colony of the white stoney coral, *Lophelia pertusa*, in the foreground. This latter species is the main reef-building organism off Mid-Norway (Based on Hovland, 2008).

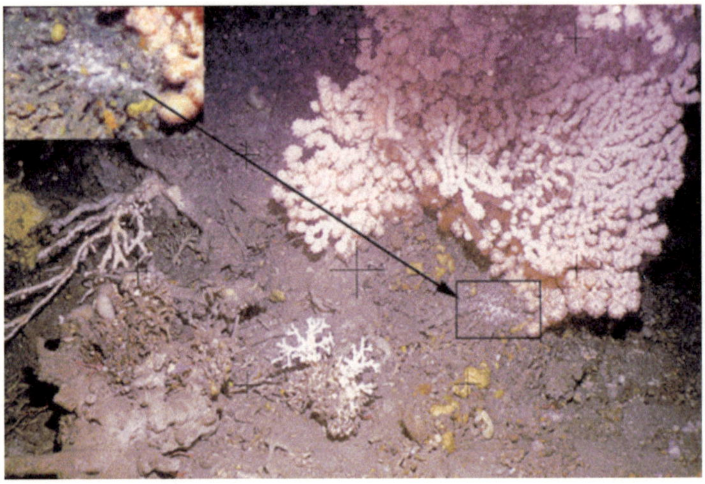

Fig. 16. A spectacular image of the seafloor, which was acquired in 1991 by use of ROV at the Onyx field (Block 6407/4, Hovland, 2008), near the seeps shown in the previous two figures. The arrowed insert image has been enlarged to show a white bacterial mat (see text for further explanation).

4.2 The deep-water coral reef (DWCR) enigma

One of the main and controversial questions still remaining to be answered with respect to the impressive deep-water coral reefs (DWCRs), found in the thousands off Mid-Norway (Hovland et al., 1995a; Mortensen et al., 1995; Hovland and Thomsen, 1997; Hovland and Risk, 2003; Hovland, 2008) is why they occur in deep and cold water, even north of the Polar Circle. This is despite there hardly being any photosynthesis occurring in the surface waters during the winter months. A 'hydraulic theory' suggesting that they rely on locally produced nutrients coming from the extra energy percolating upwards as light hydrocarbons (especially methane, ethane, and propane), was put forward by Hovland (1990b), and has been reiterated several times since then, as more supportive data occurs. However, this theory is hard to prove, and the "consensus" view among the world's leading marine biologists on this matter is still that deep-water coral reefs represent natural biologic entities that are not dependent upon such (exotic) local nutrients: "Colonial scleractinians need hard substrate for settlement. This substrate can be a shell or a pebble, and as soon as one colony is present it provides new hard substrate for subsequent colonisation." (Buhl-Mortensen et al., 2010). So, the question remains as to why only less than 0.1 per mille (‰) of the total area in the depth zones where they occur is covered by cold-water coral reefs? Why, for example, are there no more of them in the Norwegian and New Zealand fjords, where the distribution of intermediate and deep water masses is right, and where there is ample suitable hard substrate (rock bottom) with high current speeds?

However, the search for further support to the Hydraulic theory continues, and some recent microbial studies seem to point the way forward. Thus, Yakimov et al. (2006), for example, recently found metabolically active microbial communities associated with deep-water corals in the Mediterranean. Also recent research of the microbial food chain surrounding the coral reefs off Mid-Norway, have provided some interesting new findings. The contrast between coral-associated and free-living bacteria may suggest that few free-living bacteria are directly ingested by the coral and that instead, corals feed on non-bacterial plankton. Small (100 - 200 μm) zooplankton has been suggested to be important in the diet of corals (Sorokin and Sorokin, 2009). In addition, the tissue-associated bacterial communities potentially provide a direct translocation of nutrients through metabolism of particulate and dissolved organic matter in the seawater. One *Lophelia pertusa* associated bacterium was studied in more detail and named "Candidatus Mycoplasma corallicola" (Neulinger et al., 2008). It was found to be abundant in *L. pertusa* from both sides of the Atlantic Ocean and is considered an organotrophic commensalist. Given the importance of chemosynthesis in deep-water ecosystem development and functioning, cold-water coral reef communities may be linked to a diversity of chemoautotrophic microorganisms that synthesise organic compounds from inorganic compounds by extracting energy from reduced substances and by the fixation of dissolved CO_2. Just a tiny fraction of microorganisms associated with deep-water coral reefs have yet been identified, and even less assigned to a function. Although no nutritional symbiosis based on chemosynthesis (Tavormina et al., 2008) have to our knowledge been documented on deep-water coral reefs, primary producers affiliated with chemoautotrophs (utilizing H_2S, NO_2^-) and methanotrophs (utilizing CH_4) have been found associated with the reef animals and their ambient environment (Penn et al., 2006). Thus, also light hydrocarbons can probably stimulate the growth and the high biodiversity found on the *Lophelia*-reefs associated with some Norwegian hydrocarbon fields (Hovland, 2008). Only further detailed studies of the reefs will be able to answer these important questions (Jensen et al., 2008).

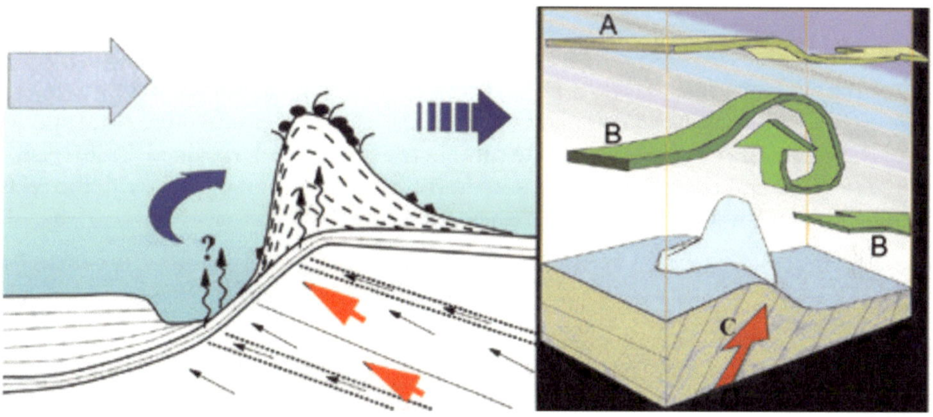

Fig. 17. The DWCR Hydraulic theory is illustrated in these two sketches. The concept is based on the idea that nutrients are brought up to the seafloor surface through seepage (red arrows). The arrow A represents the general prevailing current, B is the turbulent near-bottom current, and C, fluid flow (seepage) from below (Based on Hovland, 2008).

Despite the fact that no nutritional symbiosis involving chemosynthesis have been documented on deep-water coral reefs, other than 'unassigned' primary producers affiliated with chemoautotrophs and methanotrophs utilizing H_2S, NO_2^-, CH_4, and possibly iodide (Penn et al., 2006; Yakimov et al., 2006; Jensen et al., 2008), it was recently concluded that hydrocarbons probably stimulate the growth, the high biodiversity, and biodensity, including the rare purple octocoral found on the pockmark-*Lophelia* reef, MRR (Morvin Reference Reef), at Morvin (see Hovland, 2008, and Fig. 19). Further support to the

Hydraulic theory comes from microbial studies conducted on the large sub-surface hydrocarbon plume emitted during the Deepwater Horizon blowout in the Gulf of Mexico, in 2010 (Redmond and Valentine, 2011). They found that three types of microbial communities bloomed as a result of the hydrocarbons in the water: *Oceanospirillales, Colwellia,* and *Cycloclasticus*. The two first of these community types, which apparently utilize ethane and propane in the water, are also found in the near-bottom water surrounding DWCRs at Morvin off Mid-Norway (Jensen et al., 2008; 2010).

Fig. 18. A typical Norwegian *Lophelia*-DWCR consists of a large mound of live and dead *Lophelia pertusa* skeletons, other colonizing organisms, are sponges and other coral species. A live *Lophelia*-colony is seen here (the white coral in the foreground, Hovland, 2008). Fish, like the seithe (*Pollack*) and red fish (*Sebastes*) seen here, are normally associated with the reefs. The seithe is about 1 m long and the red fish about 30 cm.

Fig. 19. Left image, the *Sebastes* red-fish probably preparing for spawning at a *Lophelia*-reef at the Morvin field off Mid-Norway (the nearest red fish is about 35 cm long). Right, the beautiful violet coral *Anthothela grandiflora* covering a dead *Lophelia*-colony at the Morvin Reference Reef. The seithe swimming by is about 1 m long (see also Hovland, 2008).

4.3 The Fauna reef and Unit-pockmarks

A relatively large, composite (ca. 500 m long, 100 m wide, and 25 m high) DWCR, was recently discovered in association with numerous Unit-pockmarks (Hovland et al., 2010). This "Fauna reef" was named after one of the vessels used during the investigations, the "Edda Fauna". The area is located at 07° 53′ E and 63° 54′ N, near the well-known Sula Reef Complex (Freiwald et al., 2002; Hovland, 2008), off Mid-Norway (Fig. 20).

A total of 233 Unit-pockmarks and the large Fauna reef occur within the detailed surveyed rectangular area measuring 920 m by 245 m, mapped with ROV-mounted MBE (0.2m by 0.2 m gridding). Whereas 79% of the Unit-pockmarks were evenly scattered up-stream of the prevailing current at the Fauna reef (shown in Fig. 20), the rest of them were scattered up-stream of a much smaller DWCR. This latter one (ca. 50 m long, 30 m wide, 7 m high) was, however, located on the outer rim of a large Normal-pockmark crater of 180 m diameter and 16 m depth.

Previous seafloor investigations in this region have revealed DWCRs closely associated with Normal-pockmarks of up to 200 m width and 12 m depth. Such pockmark-related reefs occur at the nearby Haltenpipe Reef Cluster (HRC, Hovland and Risk, 2003; Hovland and Thomsen, 1997) and at the Kristin and Morvin hydrocarbon fields (Hovland, 2008). At all these other locations, Unit-pockmarks also occur adjacent to the reefs, and are densest up-stream with respect to the prevailing current direction (Fig. 20). Generally, the concentration of light hydrocarbons (methane - pentane) within the sediments of Unit-pockmarks is found to be higher than in the background, surrounding sediments (Hovland et al., 2010). It was concluded that Unit-pockmarks probably provide the necessary seep-related nutrients to stimulate year-round healthy coral growth and reef development, even in the dark winter season. A possible test to this theory would be to attempt detection of stable carbon-isotope variations in DWCR skeletons, as a function of season.

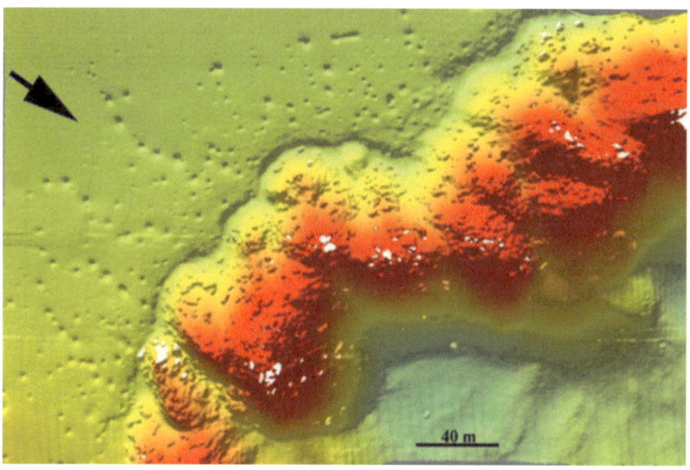

Fig. 20. The enigmatic 'Fauna Reef' off Mid-Norway. The black arrow indicates the prevailing current direction. The reef is about 25 m high (indicated here in red and yellow colours) and consists mainly of living *Lophelia pertusa* colonies. Notice the high density of Unit-pockmarks occurring in an up-stream direction relative to the reef (Based on Hovland et al., 2010). Note, the white patches are gaps in data (gaps in the DTM grid), as it is very difficult to cover the whole area of this reef-associated complex topography with ROV-MBE.

5. Other seep-related seafloor features, worldwide

In this last section, I will touch upon some other important seep-related seafloor geomorphology features that have been documented around the world over the last 30 years or so. Although the geomorphology associated with seafloor gas hydrates is very important, there is only space to cover a small portion of this issue. Three other aspects of seep/vent-associated seafloor geomorphology have also evaded detailing in this chapter. These are: 1) mud volcanism, which is perhaps even more common on the Second Surface than on land (e.g. Milkov, 2002). 2) The strange asphalt volcanoes, which tend to be associated somehow with the formation of salt domes (Hovland et al., 2005). 3) The vast subject on hydrothermal venting along tectonically active zones. Some of the pertinent information has, however, been treated by Judd and Hovland (2007).

5.1 Sandwave pattern modified by seepage

Sandwaves represent common geomorphology features on the seafloor where there is abundant sand in combination with high current velocities. Such conditions occur in the Southern North Sea and along the eastern coast of UK. During mapping campaigns offshore Belgium and the Netherlands in parts of the southern North Sea, Statoil mapped large fields of sandwaves. The waves were up to 12 m high and nearly 40 metres long. The detailed mapping was performed in association with pipeline route surveys for the laying of trunk pipelines from Norway to Belgium and France in the mid 1990's. Hull-mounted MBEs were used in combination with towed SBPs. By careful inspection of the results, it was disclosed that pockmarks had formed even in this high-energy and high-relief seafloor 'landscape' (Fig. 21). This part of the sandwave field is located where there are strong indications on the SBP-data of gas-charged sediments and peat deposits underneath the seafloor (Judd and Hovland, 2007). Slight hydroacoustic flares were also found on the original SBP-recordings, showing active ebullition through the seafloor at one of the pockmark locations.

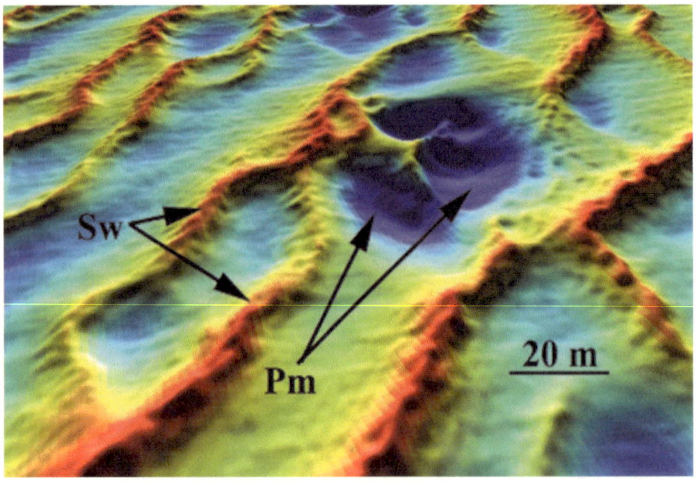

Fig. 21. A digital terrain model of a sandwave-field in the Southern North Sea. The regular sandwave pattern has clearly been disturbed by pockmark development (see text for further details). Sw=Sandwave, Pm=Pockmark. The highest sandwave crest is about 12 m above the 'normal' seafloor (greenish hue) and the deepest pockmark is about 2 m below that same depth (Based on Judd and Hovland, 2007).

The disturbance of the sandwave pattern occurs as a consequence of the 'competing' force of vertical fluid flow. This flow makes sand-grains more easily moved by the horizontal tidal currents, such that they are swept away and pile up where there is no vertical fluid flow. Thus, the sand piles up at the nearest location where this is possible, and large, sometimes pyramidal "freak sandwaves" result, thus breaching the otherwise rhythmic and regular sandwave pattern (Judd and Hovland, 2007).

5.2 Deep-sea carbonate mounds

Large, mounded structures occurring in relatively deep-water, often located inside depressions on the seafloor were first found off the SW coast of Ireland, in the Porcupine Basin (Fig. 22, Fig. 1). They were detected on 2D-seismic records and were later published by Hovland et al. (1994b). It was speculated that these structure, and similar ones, occurring on the other side of the planet, off NW Australia actually represented carbonate knolls, that had formed due to seepage, as suggested by Hovland (1990b).

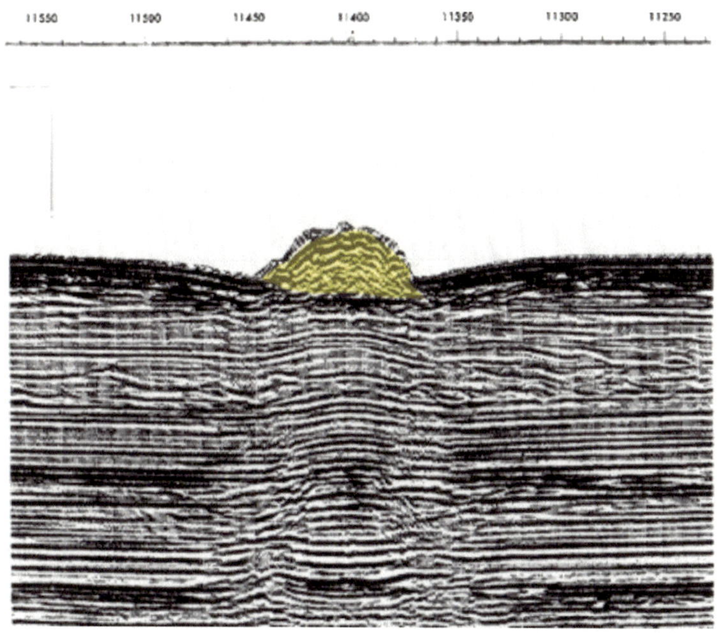

Fig. 22. A 2D reflection seismic image of a live, giant carbonate mound (yellow) off SW-Ireland (Hovland et al., 1994b). The height of the carbonate mound is about 100 m and its length at base is about 1 km. The water depth is about 800 m (see also Hovland, 2008).

The very same year as the paper came out, in 1994, at least three academic research vessels conducted further seafloor mapping offshore west and southwest Ireland. It was documented that the mounds were built up as carbonate structures, capped with living deep-water corals. However, it seems that the reason why they build up where they do, may be more complex than previously suspected, even despite drilling campaigns where one of the now dead mounds was sampled by drilling (Ferdelman et al., 2006; Hovland, 2008). Therefore, much more scientific multidisciplinary work is needed to be done before we know why and how they build up (Foucher et al., 2009).

5.3 Gas hydrates in deep-sea sediments

Gas hydrates are crystalline, ice-like compounds composed of water and gas molecules, where the small gas molecules are trapped within a cage-like framework of hydrogen-bonded water molecules (Hovland, 2005). These structures (chemically called 'clathrates') are formed under very specific temperature and pressure conditions in an environment with adequate water and gas. The right conditions can be found on land, in polar regions, where surface temperatures are very low, and within water or seabed sediments at depths exceeding about 300 m and where the temperature is adequately low (generally $< 10^oC$). The four conditions that need to be fulfilled before gas hydrates, a substance resembling ice, form are: i) The presence of abundant water (H_2O); ii) free gas molecules, small enough to fit into the water cages, i.e., methane, ethane, propane, n-butane, and/or carbon dioxide (CO_2) and hydrogen sulphide (H_2S); iii) adequately high pressure, and iv) adequately low temperature. The most common hydrate occurring in the oceanic sediments, are methane hydrates, where one cubic metre of pure methane hydrate can contain up to 169 Sm^3 (Standard cubic metres) of methane. Thus, oceanic hydrates may represent a vast energy resource currently being studied by several energy-hungry countries, including Japan, Germany, Korea, USA, Canada, and India.

Lake Baikal, in Siberia, Russia, is volumetrically the largest lake on Earth, containing about 20% of the surface fresh water volume. It was formed by rifting of the crust, and is still tectonically active, with several types of seepage phenomena, including petroleum seeps, gas seeps, and warm, hydrothermal vents (Fig. 23). After 4 years of intensive mapping of the lake, with multi-beam echosounding, coring, biological sampling, and, not least, repeated deployment of two manned Russian MIR submersibles, a series of spectacular discoveries have been made (Leifer et al., 2011):

- Several bitumen mounds, the largest being about 50 m high and still actively seeping oil,
- numerous locations of mud volcano-like mounds with gas hydrates and active gas seepage (Fig. 23), and
- new seep-related micro- and macro-animal species.

Fig. 23. Sampling gas hydrates, Lake Baikal, 2010. The image on the right is a screen shot of the on-board echosounder (EM400), showing the gas seepage from under-ground gas hydrates on a slight ridge on the lake floor (depth: 403 m below the central seep plume)

The new species discovered by sampling and visual inspection include nematodes, sponges, oligochaetes, molluscs, crustaceans, and giant flatworms. At least 10 animals are new species for science, among them two novel nematode species. Diverse microbial communities have been discovered both inside and outside bitumen structures. They consist of methanotrophic bacteria, fungi of the genus *Phitium*, eubacteria, and archaea. In the sediments near zones of naturally seeping oil, single colourless sulphur bacteria of the genus *Thioploca* were observed (Leifer et al., 2011).

The gas hydrates in Lake Baikal contain both methane and carbonate dioxide (Figs. 23 and 24) (see also De Batist et al., 2009, for more information about Lake Baikal hydrates and the associated geomorphology).

Fig. 24. The sampling of gas hydrates in Lake Baikal, 2010, was achieved by the use of a 800 kg heavy gravity corer, seen on the left being hoisted on board after having penetrated the seafloor, at 403 m water depth. After the clay-rich sediment core was split in two the result is seen on the right. There are veins and layers of pure gas hydrates inside the soft sediments.

Gas hydrates may dissociate ("melt") rapidly, when disturbed by heating on the seafloor or depressurization by lifting. When sediment-hosted hydrates dissociate they return free gas and water to the environment, which may result in a total change of the physical sediment conditions and even of the seabed topography. Consequently, at water depths and sediment depths within the gas hydrate stability zone (GHSZ), any planned interaction with the seabed has to consider the possible existence and potential formation of gas hydrates (Hovland and Gudmestad, 2001).

If there are high-porosity sediments present, such as buried sands and gravels in regions with hydrocarbon flux, within the GHSZ, these sediment layers are likely to develop into massive gas hydrate reservoirs, as demonstrated by the Mallik gas hydrate research well in the MacKenzie Delta, Canada (see for example, Judd and Hovland, 2007 and also Clennell et al., 1999). Also, with increasing hydrocarbon flux and increasing tectonic activity, the amount of gas hydrates within the near-surface sediments increases. In addition, the seafloor topography is affected, and the presence of gas hydrates may even lead to large slope failures and seafloor collapse events.

On the continental slope off Mid-Norway, at Nyegga (the north-eastern boundary of the Storegga slide scarp), there are some large, complex pockmarks with carbonate ridges inside them, Fig. 25 (Hovland et al., 2006). The water depth ranges between 600 and 800 m and the upper sediments consist of soft, sandy and silty clay. The complex pockmarks and other fluid flow features are associated with a bottom simulating reflector (BSR), and several manifestations of gas migration at depth, including vertical conduits (pipes) seen on 2D- and 3D-reflection seismic records (Ivanov et al., 2007; Plaza-Faverola et al., 2010). There are also some distinct organic-rich sediment mounds, 'hydrate pingoes', up to 1 m high and 4 m wide (Hovland and Svensen, 2006; Hovland, 2008). Work by Ivanov et al. (2007), proved that the pockmarks contain nodules or layers of gas hydrate occurring 1 to 1.5 m below surface. This discovery has provided confirmation of their status as active methane seeps or vents (Fig. 25, right image). Unit-pockmarks were also found on the seafloor outside the complex pockmarks (Hovland et al., 2010).

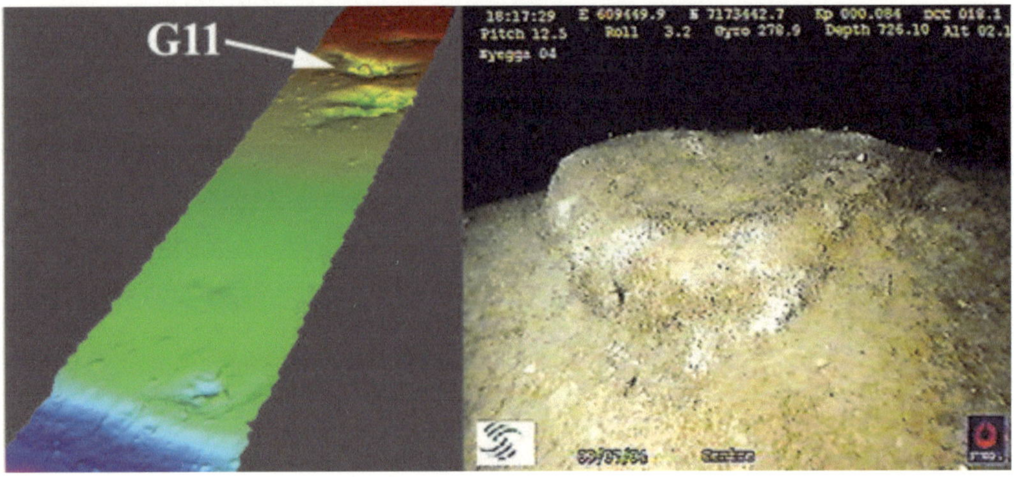

Fig. 25. At Nyegga, off Mid-Norway, there are numerous geomorphology features on the seafloor suggesting the presence of seeping gas and formation of gas hydrates under ground. In the left image, which covers a ca 6 km long and 2 km wide portion of the continental slope, there are at least three "complex pockmarks", one of which is arrowed here as "G11". Inside this pockmark there are several gas hydrate cored "pingoes", one of which is seen in the right image, its size is about 1.2 m wide (From Hovland and Svensen, 2006).

6. Conclusions

Seepage and venting of fluids through the seafloor occurs at all depths in the ocean and the processes act on the Second Surface in various ways causing local alterations in topographical expression. The most common seepage induced features found on the seafloor, are pockmark craters, which have been documented in all of the seas and oceans of our planet and also in some of the lakes. Seepage of 'exotic' (allocthonous) fluids through the seafloor not only affects the seafloor geomorphology, but also the organic (primary) production within the seafloor surface and the above water column. Hydroacoustic flares testify that the fluids spread high into the water column, and probably also affects life in the ocean to a much greater extent than previously suspected. The cross-disciplinary study of the effects of seepage and venting on the ocean and lake sediment surface and water columns can be regarded as a rapidly emerging new scientific theme.

7. References

Belka, Z. (1998). Early Devonian Kess-Kess carbonate mud mounds of the Eastern Anti-Atlas (Morocco), and their relation to submarine hydrothermal venting. J. Sed. Res., 68, 368-377.

Brooks, J.M., Bright, T.J., Bernard, B.B. & Schwab, C.R. (1979). Chemical aspects of a brine pool at the East Flower Garden bank, northwestern Gulf of Mexico. Limnology and Oceanography 24, 735-745.

Buhl-Mortensen, L., Vanreusel, A., Gooday, A.J., Levin, L.A., Priede, I.G., Buhl-Mortensen, P., Gheerardyn, H., King, N.J. & Raes, M. (2010). Biological structures as a source of habit heterogeneity and biodiversity on the deep ocean margins. Marine Ecology 31, 21-50.

Cathles, L.M., Su, Z. & Chen, D. (2010). The physics of gas chimney and pockmark formation, with implications for assessment of seafloor hazards and gas sequestration. *Marine and Petroleum Geology* 27, 82-91.

Clennell, M.B., Hovland, M., Booth, J.S., Henry, P. & Winters, W.J. (1999). Formation of natural gas hydrates in marine sediments. Part 1: Conceptual model of gas hydrate growth conditioned by host sediment properties, *J. Geophys. Res.* 104, B 10, 22985-23003.

Dando, P. & Hovland, M. (1992). Environmental effects of submarine seeping natural gas. Cont. Shelf Res. 12 (10), 1197-1207.

De Batist, M., Klerkx, J., Vanneste, M., et al. (2009). Tectonically induced gas hydrate destabilization and gas venting in Lake Baikal, Siberia. In: Abstracts of Sixth Int. conf. Gas in Marine Sediments. St Petersburg, VNIIOkeangeologia, 22-23.

Ferdelman, T.G., Kano, A., Williams, T., Henriet, J.-P. & the Expedition 307 Scientists (2006). Expedition 307 summary. Proceedings of the Integrated Ocean Drilling Program, Vol. 307, doi: 10.2204/iodp.proc.307.101.2006

Forsberg, C.F., Planke, S., Tjelta, T.I., Svanø, G., Strout, J.M. & Svensen, H. (2007). Formation of pockmarks in the Norwegian Channel. Proceedings of the 6th International Offshore Site Investiagation and Geotechnics Conference, SUT-OSIG. London, UK., 221-230.

Foucher, J-P., Westbrook, G.K., Boetius, A., Ceramicola, S., Dupré, S., Mascle, J., Mienert, J., Pfannkuche, O., Pierre, C. & Praeg, D. (2009). Structures and drivers of cold seep ecosystems.Oceanography V.22 (1), 92-109.

Freiwald, A., Hühnerback, V., Lindberg, B., Wilson, J.B. & Campbell, J. (2002). The Sula Reef Complex, Norwegian Shelf. *Facies* 47, 179-200.

Hovland, M. (1981). Characteristics of pockmarks in the Norwegian Trench. Marine Geology, 39, 103-117.

Hovland, M. (1982). Pockmarks and the Recent geology of the central section of the Norwegian Trench. Mar. Geol., 47, 283-301.

Hovland, M. (1983). Elongated depressions associated with pockmarks in the Western Slope of the Norwegian Trench. Mar Geol 51, 35-46.

Hovland, M. (1984). Gas-induced erosion features in the North Sea. Earth Surf Proc. and Landforms, 9, 209-228.

Hovland, M. (1990a). Suspected gas-associated clay diapirism on the seabed off Mid Norway. Mar. and Petrol. Geol., 7, 267-276

Hovland, M. (1990b). Do carbonate reefs form due to fluid seepage? Terra Nova, 2, 8-18.

Hovland, M. (2002). On the self-sealing nature of marine seeps. Cont Shelf Res 22: 2387-2394.

Hovland, M. (2005). Gas hydrates. In: Encyclopedia of Geology. Selley, R.C., Cocks, L.R.M. and Plimer, I.R. (eds.) Elsevier, Oxford, V. 4, 261-268.

Hovland, M. (2007). Discovery of prolific natural methane seeps at Gullfaks, northern North Sea. Geo-Marine Letters, DOI 10.1007/s00367-007-0070-6.

Hovland, M. (2008). Deep-water coral reefs: Unique Biodiversity hotspots. Praxis Publishing (Springer), Chichester, UK, 278 pp.

Hovland, M., Croker, P. & Martin, M. (1994b). Fault-associated seabed mounds (carbonate knolls?) off western Ireland and north-west Australia. Marine and Petroleum Geology 11, 232-246.

Hovland, M., Farestveit, R. & Mortensen, P.B. (1994a). Large cold-water coral reefs off mid-Norway – a problem for pipelaying? Proc. Oceanology Int., Brighton, 3, 35-40.

Hovland, M. & Gudmestad, O.T. (2001). Potential influence of gas hydrates on seabed installations. In: Paull, C.K., Dillon, W.P. (eds.) Natural gas hydrates, Occurrence, distribution, and detection. Geophysical Monograph 124, Am.Geophys. Union., 307-315.

Hovland, M., Heggland, R., de Vries, M.H.& Tjelta, T.I. (2010). Unit-pockmarks and their potential significance for prediction of fluid flow. *J. Marine and Petroleum Geol.* 27, 1190-1199.

Hovland, M., MacDonald, I., Rueslåtten, H., Johnsen, H.K., Naehr, T. & Bohrmann, G. (2005). Chapopote asphalt volcano may have been generated by supercritical water. EOS, 86 (42), 397, 402.

Hovland, M. & Risk, M. (2003). Do Norwegian deep-water coral reefs rely on seeping fluids? *Marine Geology* 198, 83-96.

Hovland, M. & Thomsen, E. (1997). Cold-water corals - are they hydrocarbon seep related?. *Marine Geology* 137, 159-164.

Hovland M. & Judd A.G. (1988) Seabed pockmarks and seepages – impact on geology, biology and the marine environment. Graham & Trotman Ltd, London, pp 293

Hovland M., Judd, A.G. & Burke Jr R.A. (1993). The global flux of methane from shallow submarine sediments. Chemosphere 26 (1-4): 559-578.

Hovland, M., Judd, A. G., King & L. H. (1984). Characteristic features of pockmarks on the North Sea Floor and Scotian Shelf. Sedimentology, 31,471-480.

Hovland M, Judd A, Maisey G (1985) North Sea gas feeds the North Sea fisheries. New Scientist 107 (1468): 26

Hovland M. & Sommerville, J. H. (1985) Characteristics of two natural gas seepages in the North Sea. Mar Petrol Geol 2: 319-326

Hovland, M., Talbot, M., Olaussen, S. & Aasberg, L. (1985) Recently formed methane-derived carbonates from the North Sea floor. In: Thomas BM (Ed.) Petroleum Geochemistry in Exploration of the Norwegian Shelf.

Ivanov, M., Westbrook, G.K., Blinova, V., Kozlova, E., Mazzini, A., Nouzé, H. & Minshull, T.A. (2007). First sampling of gas hydrates from the Vøring Plateau. Eos Transaction AGU 88 (19), 200 & 212.

Jensen, S., Neufeld, J.D., Birkeland, N.-K., Hovland, M. & Murrell, J.C. (2008). Methane assimilation by *Methylomicrobium* in deep-water coral reef sediment off the coast of Norway. Submitted to: FEMS, Microbiology Ecology 66, 320-330.

Jensen, S., Neufeld, J.D., Birkeland, N.-K. & Hovland, M. (2010). Intracellular *Oceanospirillales* bacteria inhabit gills of *Acesta* bivalves. FEMS, Microbiology Ecology 74, 523-533.

Judd, A. & Hovland, M. (2007). Seabed fluid flow - impact on geology, biology and the marine environment. Cambridge University Press, Cambridge, pp 400. www.cambridge.org

King, L.H. & MacLean, B. (1970). Pockmarks on the Scotian Shelf. Geol Soc Am Bull 81: 3142-3148.

Leifer, I., Hovland, M., Zemskaya, T., 2011. Two decades of community research on gas in shallow marine sediments, EOS (AGU) 92 (15), 128.

Mortensen, P.B., Hovland, M., Brattegard, T. & Farestveit, R. (1995). Deep water bioherms of the scleractinian coral Lophelia pertusa (L.) at 64oN on the Norwegian shelf: structure and associated megafauna. Sarsia 80, 145-158.

Milkov, A.V. (2000). Worldwide distribution of submarine mud volcanoes and associated gas hydrates. Marine Geol. 167, 29-42.

Nelson, D.C. & Fisher, C.R. (1995). Chemoautotrophic and methanotrophic endosymbiotic bacteria at deep-sea vents and seeps. In: Karl, D.M. (ed.), The Microbiology of Deep-Sea Hydrothermal Vents. CRC Press, Boca Raton, 125-167.

Neulinger, S.C., Gärtner, A., Järnegren, J., Ludvigsen, M., Lochte, K. & Dullo, W.-C. (2008). Tissue-associated "*Candidatus Mycoplasma corallicola*" and filamentous bacteria on the cold-water coral *Lophelia pertusa* (Schleractinia). Applied and Environmental Microbiology 75, 1437-1444.

Niemann, H., Elvert, M., Hovland, M., Orcutt, B., Judd, A.G., Suck, I., Gutt, J., Joye, S., Damm, E., Finster, K. & Boetius, A. (2005). Methane emission and consumption at a North Sea gas seep (Tommeliten area). Biogeosciences 2, 335-351.

Penn, K., Wu, D., Eisen, J.A. & Ward, N. (2006). Characteristics of bacterial communities associated with deep-sea corals on Gulf of Alaska seamounts. Applied Environmental Microbiology 72, 1680-1683.

Plaza-Faverola, A., Bünz, S. & Mienert, J. (2010). Fluid distributions inferred from P-wave velocity and reflection seismic amplitude anomalies beneath the Nyegga pockmark field of the mid-Norwegian margin. Marine and Petroleum Geology 27, 46-60.

Redmond, M.C. & Valentine, D.L. (2011). Natural gas and temperature structured a microbial community response to the *Deepwater Horizon* oil spill. PNAS, doi/10.1073/pnas.1108756108.

Reeburg, W.S. (2007). Oceanic methane biogeochemistry. Chem. Rev.107, 486-513.

Schneider von Deimling, J., Greinert, J., Chapman, N.R., Rbbel, W. & Linke, P. (2010). Acoustic imaging of natural gas seepage in the North Sea: sensing bubbles controlled by variable currents. Limnol. Oceanorgr: Methods 8, 155-171.

Sorokin, Yu.I. & Sorokin Yu. P. (2009). Analysis of plankton in the southern Great Barrier Reef: abundance and roles in throphodynamics. J. Mar. Biol. Assoc. UK 89, 235-241.

Stein, J.L. (1984). Subtidal gastropods consume sulphur-oxidizing bacteria: evidence from coastal hydrothermal vents. Science 233, 696-698.

Tavormina, P.L., Ussler, W. & Orphan, V.J. (2008). Planktonic and sediment-associated aerobic methanotrophs in two seep systems along the North American margin. Appl. Environ. Microbiol. 74, 3985-3995.

Tjelta, T.I., Svanø, G., Strout, J.M., Forsberg, C.F., Johansen, H. & Planke, S. (2007). Shallow gas and its multiple impact on a North Sea production platform. Proceedings of the 6th Int. Offshore Site Investigation and Geotechnics Conf., 11-13 Sept., London, 205-220.

Yakimov, M.M., Cappello, S., Crisafi, E., Tursi, A., Savini, A., Corselli, C., Scarfi, S. & Giuliano, L. (2006). Phylogenic survey of metabolically active microbial communities associated with the deep-sea coral *Lophelia pertusa* from the Apulian plateau, Central Mediterranean Sea. Deep-Sea Research I 53, 62-75.

Van Weering, T.C.E., Jansen, J.H.F. & Eisma, D. (1973). Acoustic reflection profiles of the Norwegian Channel between Oslo and Bergen. Netherlands Journal of Sea Research, 6, 241-264.

Webb, K.E., Barnes, D.K.A. & Planke, S. (2010). Pockmarks: refuges for marine benthic biodiversity. Marine Geology.

Wegener, G., Shovitri, M., Knittel, K., Niemann, H., Hovland, M. & Boetius, A. (2008). Biogeochemical processes and microbial diversity of the Gullfaks and Tommeliten methane seeps (Northern North Sea). Biogeosciences 5, 1127-1144.

Wendt, J., Belka, Z., Kaufmann, B., Kostrewa, R. & Hayer, J. (1997). The World's most spectacular carbonate mounds (Middle Devonian, Algerian Sahara). J. Sed. Res. 67 (3), 424-436.

5

The Use of Digital Elevation Models (DEMs) for Bathymetry Development in Large Tropical Reservoirs

José de Anda* et al.**

Centro de Investigación y Asistencia en Tecnología y Diseño del Estado de Jalisco, A.C. Normalistas 800, Guadalajara, Jalisco México

1. Introduction

Bathymetry is the study of underwater depth of lake or ocean floors. In other words, bathymetry is the underwater equivalent to hypsometry (Miller et al., 2010), and this can be used also to describe the shape and volume of water reservoirs (Obregon et al., 2011). The bathymetry is generally obtained by recording water depths throughout a water body and connecting the recorded points of equal water depth. Hence, a bathymetric map is estimated from the water depth between two points of a known depth. There may be discrepancies in any given map depending on the number of depth measurements taken: the more depth measurements recorded, the more accurate the map is.

Important qualities such as the storage-capacity curve are derived from the bathymetry and it is also crucial to understand how a system functions, including surface area, maximum length, mean width, maximum width, mean depth, maximum depth, shoreline length and volume. The amount of detail in a bathymetry dataset depends on the resolution of the mapping system, its spatial accuracy, the amount of time and effort expended in making it, as well as consideration of its intended use (Ceyhun & Yalçın, 2010).

Most of large reservoirs are formed by a dam across the course of a river, with subsequent inundation of the upstream land surface (Xu et al., 2011). The bathymetric shape of the reservoir basin coupled with the design and operation of the dam are critical to correctly modeling water quality. The combination of climate, reservoir bathymetry, seasonal hydrology, water chemistry, local wind, and temperature are all critical to modeling a

* Corresponding Author

** Jesus Gabriel Rangel-Peraza[1], Oliver Obregon[2], James Nelson[2], Gustavious P. Williams[2], Yazmín Jarquín-Javier[1], Jerry Miller[3] and Michael Rode[4]

[1]*Centro de Investigación y Asistencia en Tecnología y Diseño del Estado de Jalisco, A.C. Normalistas 800, Guadalajara, Jalisco México*

[2]*Brigham Young University, Provo, Utah, USA*

[3]*Retired Water Quality Scientist, USA*

[4]*UFZ-Helmholtz Centre for Environmental Research, Department of Hydrological Modelling. Buckstrasse 3, Magdeburg, Germany*

reservoir sufficiently correctly to test various dam designs or operational scenarios (Manivanan, 2008). The description of the water body in the bathymetry must be correct enough to simulate the reservoir hydrodynamics. Significant errors in the bathymetric description of the water body may prevent basic hydrodynamic calibration of the model. Reservoir bathymetric characteristics are, therefore, the starting point for water quality modeling (Zhao et al., 2011).

Just because a reservoir was once dry land, the bathymetry of this type of waterbody can be relatively easier to develop than lakes; this is due to the area capacity curve and topographic data developed in planning to build a dam. However, if the reservoir is old enough to have collected significant sediment deposition then additional sonar cross-section data may still have to be collected as in a natural lake.

The main scope of this work is to develop the bathymetry of Aguamilpa reservoir using spatial information in order to generate a tool that explains the elevation-capacity relationship for one of the largest tropical reservoirs in the world located in the western part of Mexico. Reliable storage capacity data for all water uses in the reservoir and useful information for current and future water quality assessments are provided with the development of this important tool.

2. Study area

The Aguamilpa reservoir is located in the central region of the state of Nayarit, Mexico and covers parts of the municipalities of Nayar, La Yesca, Santa Maria del Oro and Tepic. The Aguamilpa concrete-faced rockfill dam with 187 m high is one of the highest on its type in the world (Rangel-Peraza et al., 2009; Ibarra-Montoya et al., 2010). The main water contributions to the reservoir come from the Santiago and Huaynamota rivers. The reservoir is approximately 60 km long following the Santiago river course and 20 km along the Huaynamota River, covering a 109 km^2 area.

The morphometric and morphological features of Aguamilpa reservoir are the result of the confluence of the Sierra Madre Occidental and Neovolcanic axis, two of the largest and most important mountainous systems in Mexico. A strike-slip fault, whose trace is now occupied by the Santiago River, produces extension and subsidence in the upper basin, while in the lower basin produces compression and uplift (INEGI, 2005).

The Aguamilpa hydroelectric dam is located inside of Santiago-Aguamilpa basin in the southwestern part of the Sierra Madre Occidental (Figure 1). This area is characterized by extrusive volcanic rocks (rhyolite-acid tuff) from the Miocene, intruded by dykes of various origins (Figures 2 and 3).

Extrusive volcanic rocks have been classified into 3 units: the lower is the Aguamilpa unit, the intermediate is named Colorines unit, and the highest is Picachos unit; the first consists of mass ignimbrite while the other two have pseudo stratification according to the Federal Electricity Commission (CFE, 1997).

The main geological structural features identified at the site, correspond to six faults oriented NE-SW, known as Colorines system. Four of these faults are located on the right bank and affect the generation works. The other two are located on the left and one of them involves the detour work and spillway. There are also four major fractures which show greater horizontal than vertical continuity (Mendes, 1995).

Like any other concrete-faced rockfill dam, Aguamilpa reservoir required a footing (or plinth) to be constructed around its upstream edge. The plinth, located in the Aguamilpa unit, was made from concrete and serves as a connection between the dam and the valley walls and floor. On the other hand, water intake and excess discharge (spillway) structures were excavated in Colorines and Picachos units, respectively (Mendes, 2005).

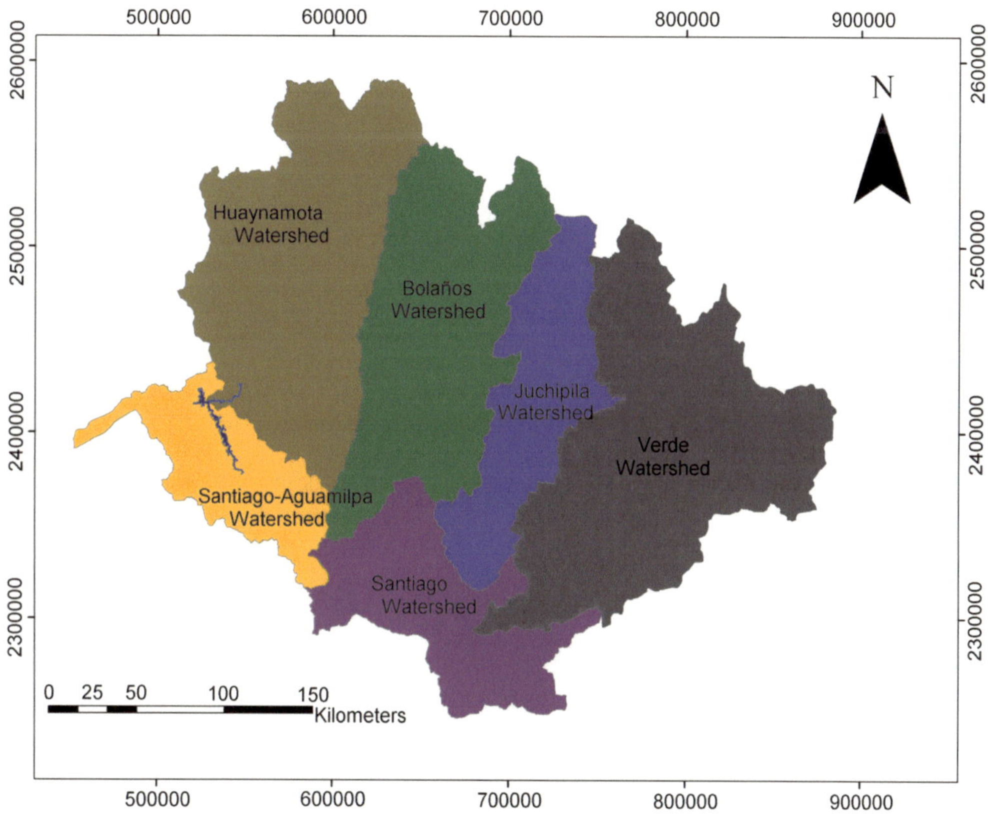

Fig. 1. Aguamilpa reservoir location in the Santiago-Aguamilpa watershed.

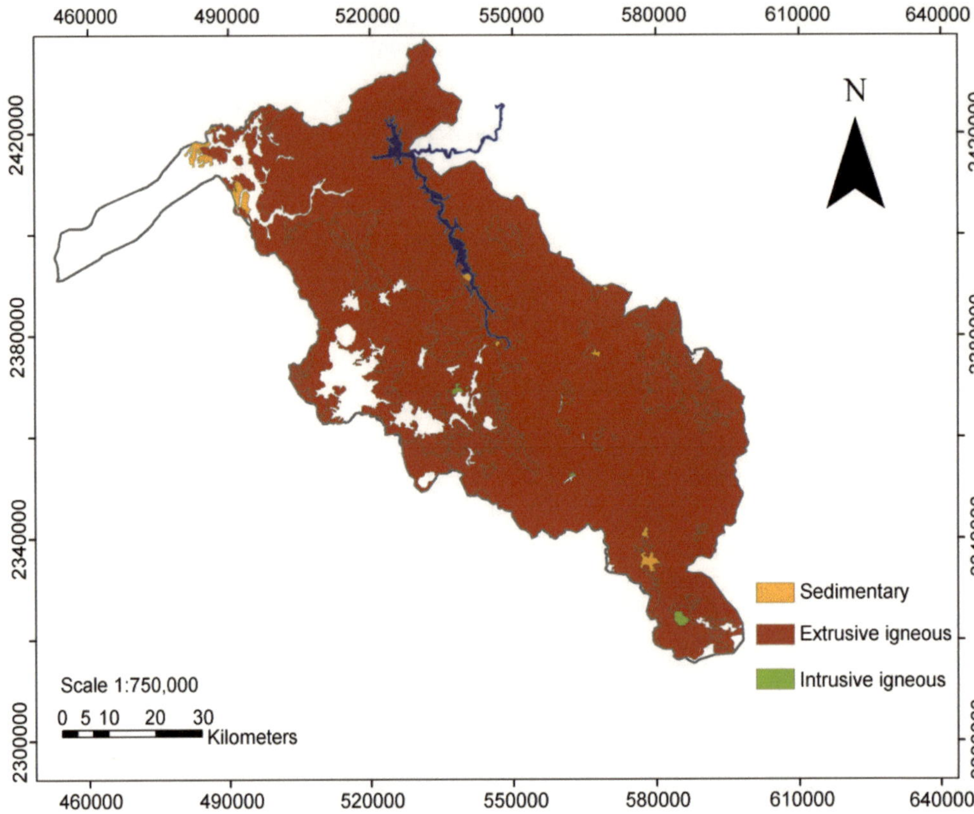

Fig. 2. Extrusive volcanic rocks distribution in the Santiago basin.

3. Materials and methods

All the aforementioned features were represented in a bathymetric map. This map represents an important tool for water quality modeling proposes due to it is highly relation to the reservoir hydrologic/hydrodynamic regime (Moses et al., 2011). Along with meteorological and hydrodynamic data and water quality parameters, it helps to determine the hydrodynamic behavior of the reservoir, improving the knowledge of vertical mixing processes, stratification and reaeration.

The development of a hydrodynamic and water quality model begins by creating the reservoir's bathymetry data. Getting a high-quality bathymetry is important to create an accurate model. The bathymetry of the Aguamilpa reservoir was created by using 1:50,000 scale Digital Elevation Models (DEMs), which were obtained from National Institute of Statistics, Geography and Informatics of Mexico (INEGI), and the Watershed Modeling System version 8.0 (WMS) software (Nelson, 2006). The necessary DEMs to cover the Aguamilpa reservoir were: F13D11, F13D12, F13D13, F13D21, F13D22, F13D23, F13D31, F13D32 and F13D33 according with the National Institute of Statistics, Geography and Informatics (INEGI, 2008). DEMs were available in the spatial resolution of 50x50 meter grids.

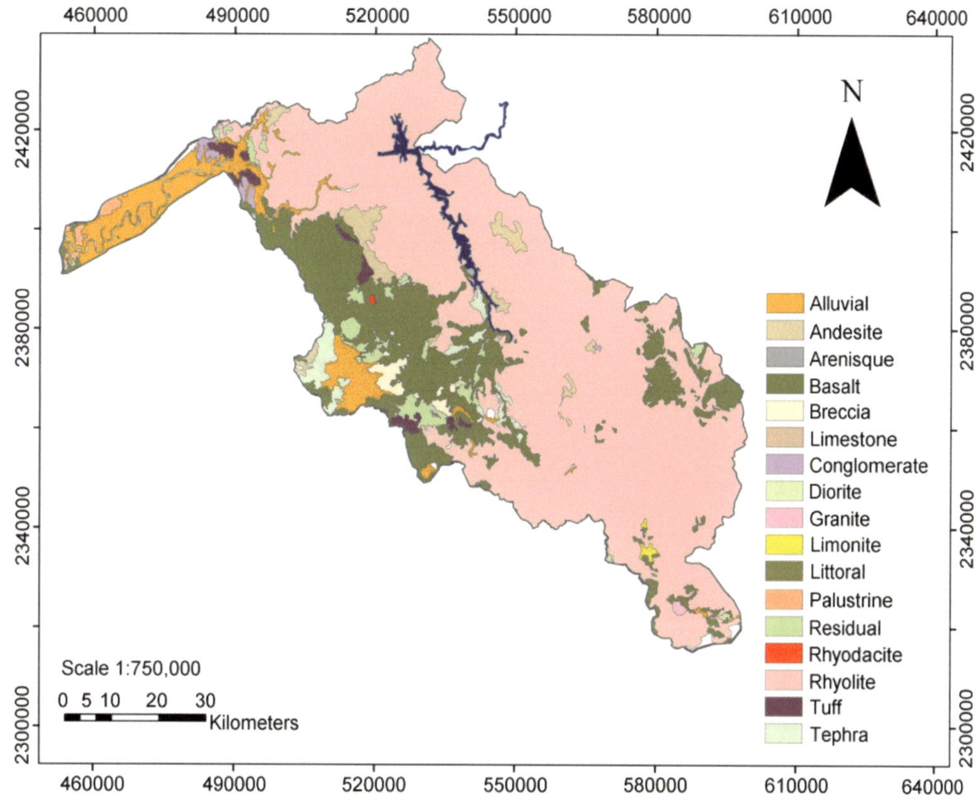

Fig. 3. Volcanic rock types distribution in the Santiago-Aguamilpa basin.

WMS is defined by Aquaveo (2011) as a comprehensive graphical modeling environment for all phases of watershed hydrology and hydraulics. WMS provides a variety of capabilities which include cross-section extraction from terrain data, watershed delineation, calculation of the geometry watershed and others. Similar to a Geographic Information System (GIS) format, spatial data was read into WMS and processed. DEMs were converted to Triangulated Irregular Networks (TINs) using WMS to define the boundaries and storage capacity of the Aguamilpa reservoir.

A pre-processor named W2i, included in the W2i-AGPM Modeling System User Interface for CE-QUAL-W2, is a powerful water quality modeling tool created and managed by Loginetics, Inc. (Hauser, 2007). W2i was used in several occasions to check the created bathymetry of the Aguamilpa reservoir. Viewing the bathymetry in W2i pre-processor allowed an improvement in the bathymetry created by WMS.

The resulting bathymetry was used to compute surface and volume curves. This information was then compared with the hypsographic curves provided by the Federal Electricity Commission (CFE, 2002) to validate the results obtained. Bathymetry accuracy was evaluated by the percentage error statistic. This statistic was calculated from the relative error which is the quotient between the absolute error and the reference value. The percentage error is 100% times the relative error.

4. Results and discussion

WMS processed the DEM data to produce a bathymetry for Aguamilpa reservoir, which is compatible with some water quality models, such as CE-QUAL W2 (Obregon et al., 2011). DEMs were converted to Triangulated Irregular Networks (TINs) and the boundaries of the reservoir were defined (Figure 4).

The reservoir's boundaries were set at a maximum elevation of 235 meters (corresponding to the high-water level in the reservoir), resulting in an extension of the reservoir approximately 60 km along the Santiago River and 25 km along the Huaynamota River (Figure 5).

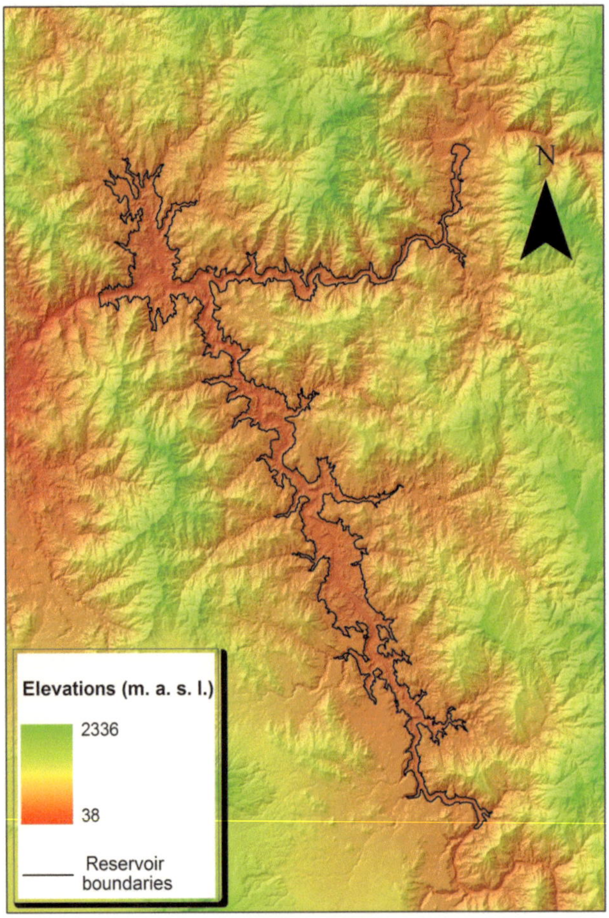

Fig. 4. DEM and boundaries of Aguamilpa reservoir.

Using this bathymetry, the Aguamilpa water body was discretized into a series of longitudinal segments. This was done by creating polygons longitudinally along the reservoir. After specifying a layer height of 1 meter for all cells, the cell widths were calculated from the TIN developed for Aguamilpa reservoir. The segments widths were calculated by using the length, depth and volume of the segments as following:

$$\text{Width} = \frac{\text{Volume}}{\text{Depth} * \text{Length}}$$

According to this, Aguamilpa reservoir was divided into segments each roughly 500 meters long. At the beginning, WMS generated a maximum number of 155 layers with one meter thickness each. This situation was related to the minimum elevation of the used DEM's, which was 80 meters. However, 13 more layers were added by hand to the 155 generated layers (1 meter) to obtain 168 layers of a maximum number for each single segment.

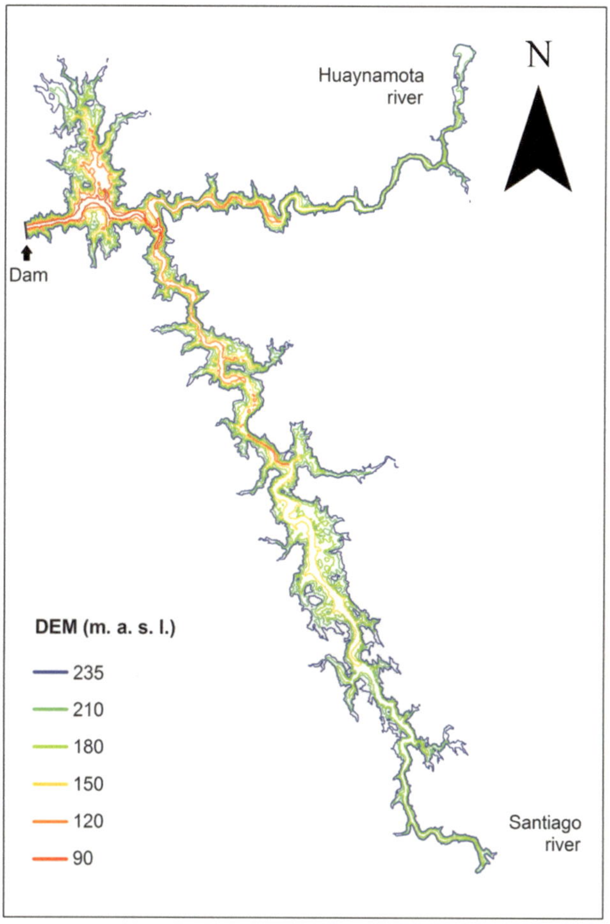

Fig. 5. Bathymetry of Aguamilpa reservoir.

This is because the maximum reservoir's depth is 187 meters not 155 meters. Another bathymetry study for Aguamilpa reservoir confirmed this information and reported that the minimum elevation is observed at 64.3 meters (GRUBA, 1997). The segment lengths and layer heights agree with those used by Ha and Lee (2007) in Daecheong Reservoir and Debele et al. (2006) in Cedar Creek Reservoir. This situation guarantees a sufficiently refined grid in the Aguamilpa bathymetry model. This is an example of an error in the bathymetry which could impact in the results of a hydrodynamic model of the reservoir and thus its calibration.

The bathymetry of the Aguamilpa reservoir includes a total of 3 branches and 103 segments along the three tributary branches and a total of 168 layers in its deepest part. The three created branches are identified as: Branch 1 (Santiago River), Branch 2 (Huaynamota River) and Branch 3 (Ensenada) with Branch 1 being the largest and Branch 3 the shortest. The average segment length and width were 1,215 meters and 1,349 meters respectively.

The largest segment length was 1,724 meters and the maximum segment width was 3,032 meters. The shortest segment measures 487 meters and the narrowest segment measures 217 meters. The final bathymetry grid of the branch 1 (Santiago river) is presented in Figure 6, and the final bathymetry grid of branch 2 (Huaynamota river) is shown in Figure 7. These bathymetric profiles were depicted using W2i preprocessor.

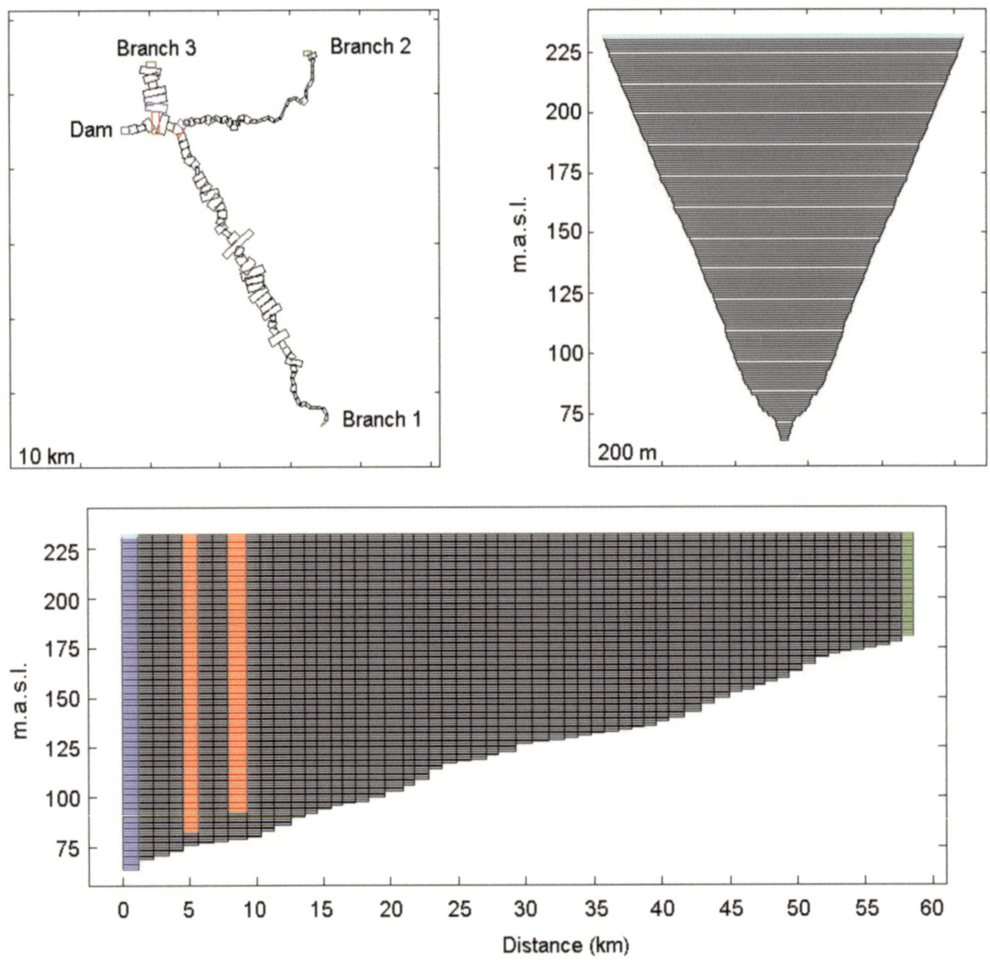

Fig. 6. Bathymety grid of Santiago river or Branch 1. This graphic illustrates the location of the aforementioned branch at the top left side and shows the cross-section of its deepest part at the top right side. The cross-section is then identified with a blue color in the horizontal profile of this branch at the bottom of this figure, where the red grids represent the connection between Branch 2 and 3 with Branch 1 and green grids correspond to the shallowest section of Santiago river.

The Aguamilpa reservoir presents an extended and narrow shape, with a typical bathymetric profile of a river-dammed reservoir. The main tributary river of Aguamilpa reservoir is located about 55 km from the wall dam.

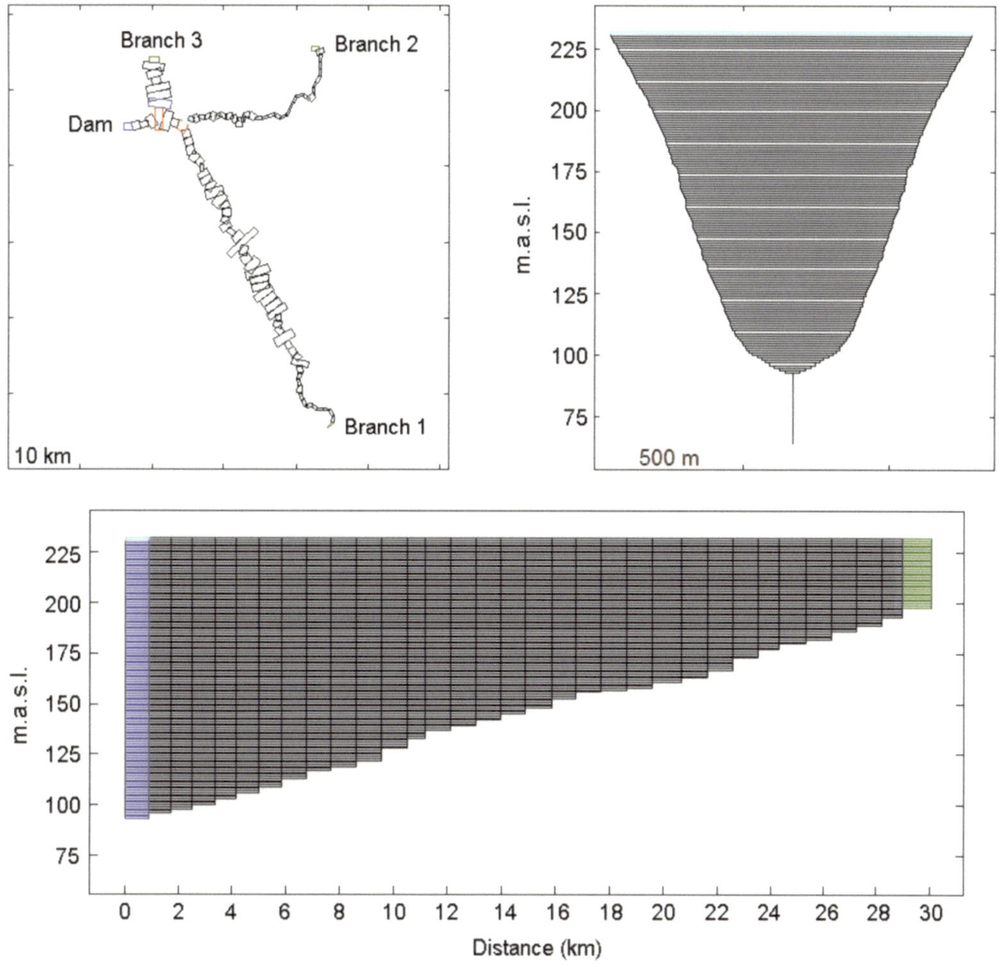

Fig. 7. Bathymetry grid of Huaynamota river or Branch 2. This graphic illustrates the location of the aforementioned branch at the top left side and shows the cross-section of its deepest part at the top right side. This cross-section is then identified with a blue color in the horizontal profile at the bottom of this figure, where the green grids correspond to the shallowest section of this river.

According to this information, it is possible to identify the presence of a longitudinal zonation in Aguamilpa reservoir: a riverine zone, a transitional zone and a lacustrine zone. The existence of these zones in the reservoir agrees with other reservoir bathymetry studies done by Margalef (1983), Wetzel (1993) and Comerma et al. (2003). This information can be used to understand some limnological conditions that could be associated to morphology in Aguamilpa reservoir.

The main morphometric features of Aguamilpa reservoir were evaluated using the results obtained in the bathymetry. These features were estimated according to methodology suggested by Hutchinson (1957), Wetzel and Likens (2000) and Torres-Orozco (2007) and are shown in Table 1.

Parameter (Units)	Value
Total volume (Mm^3)	6,933.4
Surface area (km^2)	109
Maximum lenght (km)	58
Maximum width (m)	4,952
Maximum depth (m)	164
Mean depth	63

Table 1. Morphometric features of Aguamilpa reservoir.

The maximum depth of Aguamilpa reservoir is much greater than its mean depth having a maximum depth to mean depth ratio of 3.74:1. These results demonstrate that a considerable proportion of the reservoir total volume is derived from relatively shallow areas which were created when the former riverine flood plain was permanently flooded.

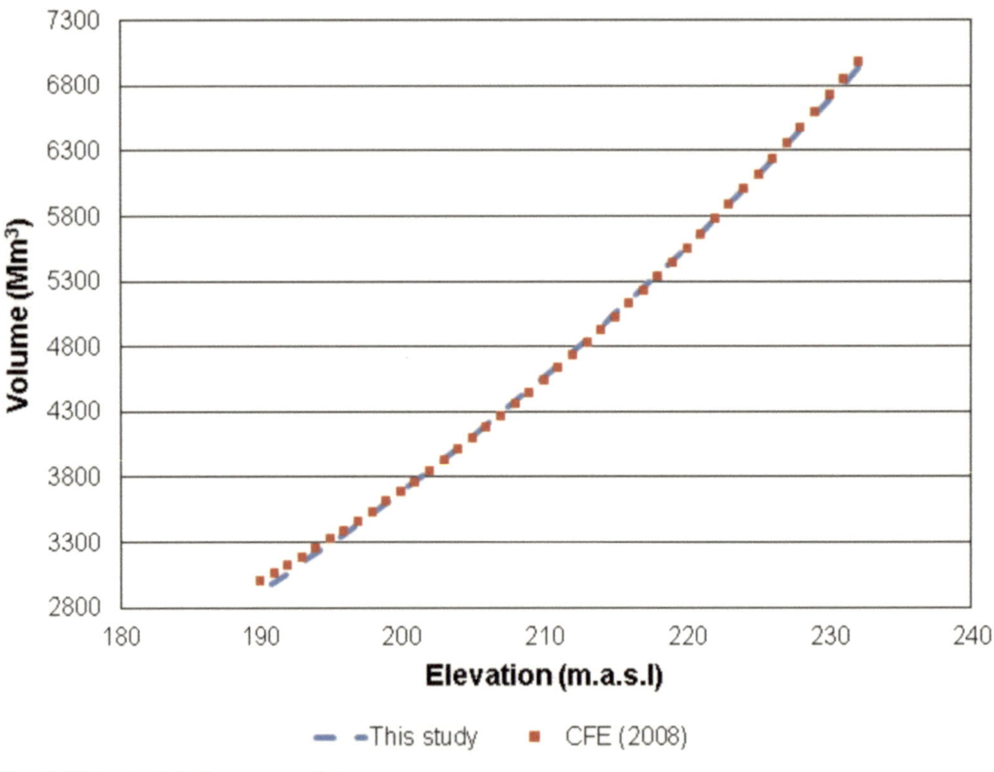

Fig. 8. This study's hypsographic curves.

A storage capacity curve was then created from the bathymetry file and was compared with the storage capacity obtained from CFE obtained to check the accuracy of the bathymetry data. As shown in Figure 8, the created storage capacity curve was similar to the curve reported by CFE.

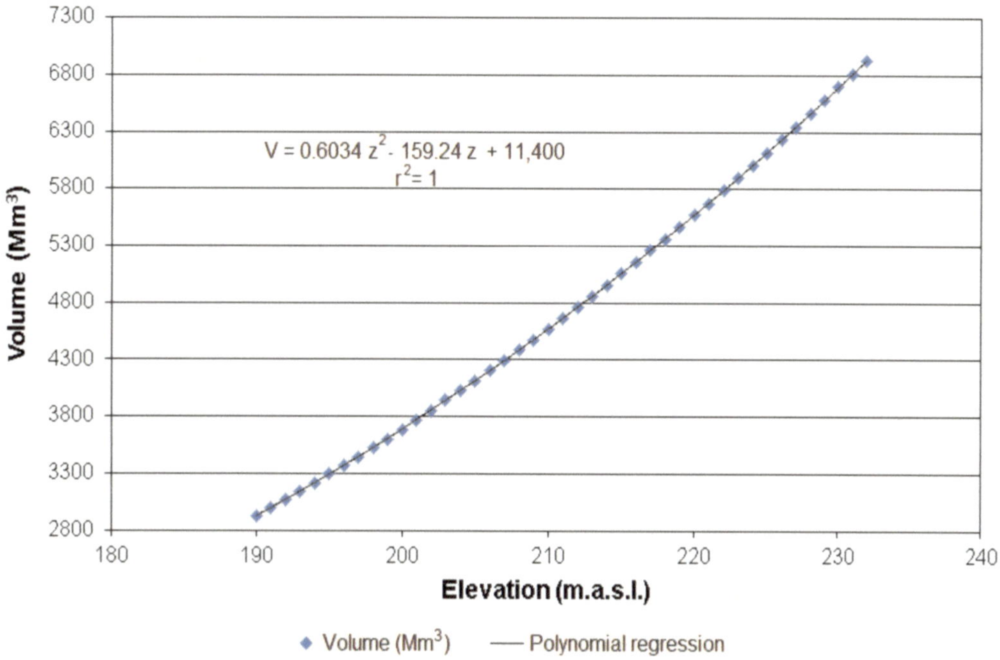

Fig. 9. Reservoir volume-elevation polynomial regression.

This comparison showed that the DEM 1:50,000 available from INEGI (2008) and used for the present study was adequate to develop an accurate bathymetry for the Aguamilpa reservoir. This situation is also demonstrated with the low relative error showed in the model. The mean relative error between the official and present bathymetry was 0.3%. Therefore, the use of DEMs simplified the process of bathymetry development for water quality modeling making forecasts based on the predetermined full supply level contour.

It is important not only to maintain the overall volume area capacity curve, but also to stay very close through each elevation zone. Since inflow will seek their own density level in the reservoir, hydrodynamic calibration can be altered by errors at critical elevations in volume and general shape of the bathymetry file. Temperature is usually the primary controller of this density placement with depth in the reservoir.

A regression was carried out on the bathymetric information relative to reservoir volume and elevation (Figure 9). The data was fitted by a second degree polynomial with the least squares method. The equation that best describes the volume-elevation ratio was:

$$V = 11{,}400 - 159.24\,z + 0.6034\,z^2$$

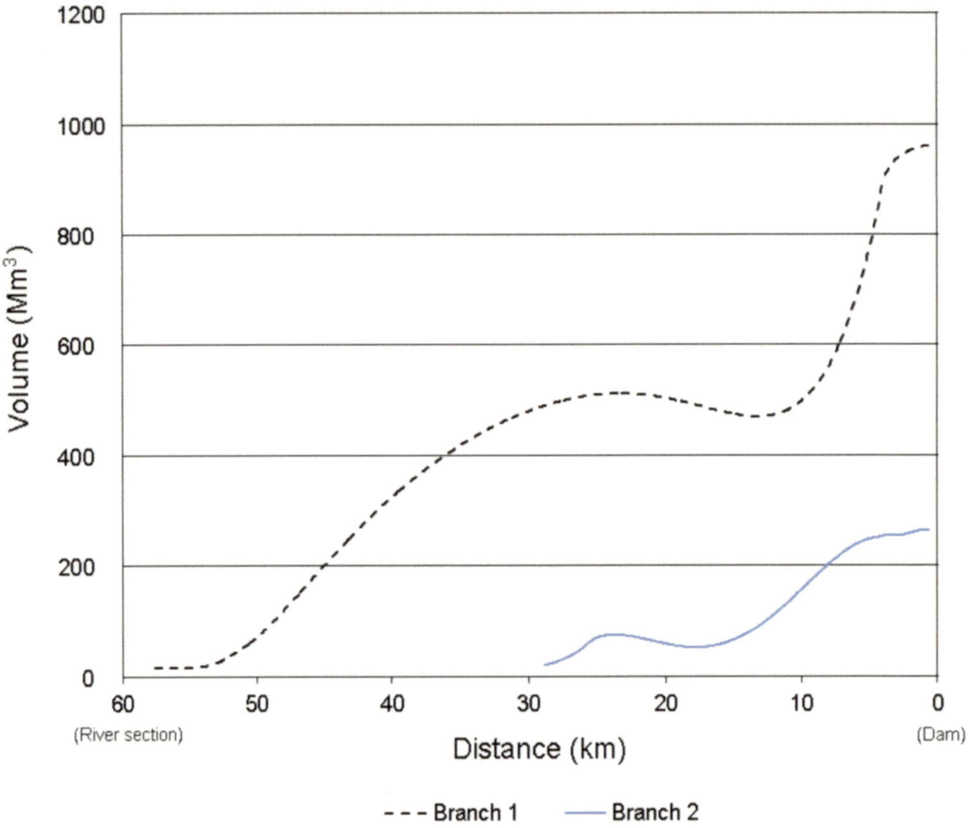

Fig. 10. Aguamilpa reservoir volume as a function of distance from the dam.

Where V is the volume in Mm^3 and z is surface height in meters above sea level (m.a.s.l.). This ratio showed a high correlation coefficient (r^2=1) for a water surface altitude range from 190 m.a.s.l. to 235 m.a.s.l, which are the operation levels of the Aguamilpa reservoir (minimum and maximum, respectively).

According to this equation, it is possible to compute reservoir volumes as a function of water elevation. This information is important for conservation and flood control of the reservoir. Storage capacity curves were also created for each branch and segment. These storage capacity curves show the storage of the reservoir at different locations (Figure 10).

5. Conclusions

A methodology was applied to demonstrate that a good resolution bathymetry can be developed by using DEM in one of the largest tropical reservoirs. A good agreement between the bathymetry computed in the present work and the hypsographic curves provided by CFE in the Aguamilpa reservoir was identified, with 0.3% of mean relative error.

Commonly scientist and engineers use the data obtained from sonar mounted beneath or over the side of a boat to make bathymetric maps. The use of WMS was advantageous because creating bathymetry for Aguamilpa reservoir took hours instead of weeks. WMS also allowed modification of the bathymetry to ensure that results were not adversely affected by grid resolution. However, if all the data must be collected using boats and sonar equipment, these surveys will still takes weeks to attain and this data collection is expensive.

The application of this methodology supports the use of DEMs for the development of reservoir bathymetries and may be applied to other reservoirs. The proposed tool would also increase stability and decrease run times in further water quality models. The use of WMS can also help quickly modify the bathymetry to simpler or more complex forms to determine if there is a critical point of complexity at which hydrodynamic results will be compromised. Ultimately using the simplest bathymetry that produces correct results can reduce model computation time.

6. Acknowledgments

The authors thank CFE for the provided elevation-capacity curve information and INEGI for their support in DEMs data. They are also thankful for the support provided by CONACYT Basic Science funds and the scholarship granted for the postgraduate studies of the first and second author.

7. References

Aquaveo. (2010). Watershed Modeling System, In: *WMS 8.4 Tutorials – Volume 6*, Accessed on January 15th 2011, Available from: http://wmstutorials.aquaveo.com/Tutor84_Vol_VI.pdf.

Ceyhun, Ö. & Yalçın, A. (2010). Remote sensing of water depths in shallow waters via artificial neural networks, *Estuarine, Coastal and Shelf Science*, 89, 89-96, ISSN: 0272-7714.

CFE. (1997). *Proyecto Hidroelectrico Aguamilpa, Nayarit Mexico,* Official report of Federal Commission of Electricity, Nayarit, Mexico, 36 p.

CFE. (2002). *Informe Final de los Estudios Hidrologicos e Hidraulicos*, Federal Commission of Electricity, Nayarit, Mexico, 40 p.

Comerma, M.; Garcia, J.C.; Romero, M.; Armengol, J. & Simek, K. (2003). Carbon Flow Dynamics in the Pelagic Community of the Sau Reservoir (Catalonia NE Spain), *Hydrobiologia*, 504, 87-98, ISSN: 0018-8158.

Debele, B.; Srinivasan, R. & Parlange, J.-Y. (2006). Coupling upland watershed and downstream waterbody hydrodynamic and water quality models (SWAT and CE QUAL-W2) for better water resources management in complex river basins, *Environmental Modeling and Assessment Journal*, 13,135-153, ISSN: 1420-2026.

GRUBA. (1997). *Estudio de Calidad del Agua en el Embalse de la Presa Aguamilpa, Nayarit,* National Commission of Water, Report Number GSCA 007/97, Nayarit, Mexico, 109 p.

Ha, S.-R. & Lee, J.-Y. (2007). Application of CE-QUAL W2 Model to Eutrophication Simulation in Daecheong Reservoir Stratified by Turbidity Storms, *Proceedings of Taal 2007-The 12th World Lake Conference*, pp. 824-833, Jaipur, India, 29th Oct - 2th Nov, 2007.

Hutchinson, G.E. (Ed.) (1957). *A treatise on limnology. Volume I: Geography, Physics and Chemistry*, John Wiley, ISBN: 9780471425700, New York, USA. 1015 p.

Hauser, G. (2007). Loginetics AGMP, CE-QUAL-W2 model post processor. In: *Thermal and Bioenergetics Modeling for Balancing Energy and Environment*, Accessed on August 30th, 2010, Available from http://www.loginetics.com/pubsm/Modeling.html.

Ibarra-Montoya, J.L., Rangel-Peraza, G., González-Farías, F.A., de Anda, J., Zamudio-Reséndiz, M.E., Martínez-Meyer, E. & Macias-Cuellar, H. (2010). Modelo de nicho ecológico para predecir la distribución potencial de fitoplancton en la Presa Hidroeléctrica Aguamilpa, Nayarit. México. *Ambi-Agua*. 5(3), 60-75, ISSN: 1980-993X.

INEGI. (2005). *Anuario Estadístico del Estado de Nayarit. Edición 2005*. National Institute of Statistics, Geography and Informatics of Mexico y Nayarit State Government. Nayarit, Mexico. 580 p.

INEGI. (2008). Sistema de Descarga del Continuo de Elevaciones Mexicano, In: *Sistema Nacional de Informacion Estadistica y Geografica*. Accessed on January 28th, 2011 Available from:
http://mapserver.inegi.org.mx/DescargaMDEWeb/?s=geo&c=977

Manivanan, R. (2008). Water quality modelling: basics, In: *Water Quality Modeling. Rivers, Streams and Estuaries*, R. Manivanan (Eds), pp. 15-28, New India Publishing Agency, ISBN: 1397889422936, New Delhi, India.

Margalef, R. (1983). *Limnologia*. Eds. Omega, ISBN: 9788428207140, Barcelona, Spain, 1024 p

Mendes, F. (1995) Aguamilpa underground penstocks - excavation phase, International *Journal of Rock Mechanics and Mining Sciences and Geomechanics Abstracts*, 32(4), 188-188, ISSN 0148-9062

Mendes, F. (2005). Rapid construction of the El Cajon CFRD, Mexico. *International Journal on Hydropower and Dams*. 12(1), 67-71, ISSN 1352-2523.

Miller, F. P., Vandome, A. F., McBrewster, J. (2010). Bathymetry. VDM Publishing House Ltd., 2010 –68 p. ISBN: 6130704542.

Moses, S.A., Janaki, L., Joseph, S., Justus, J., Vimala, S.R. (2011). Influence of lake morphology on water quality, *Environ. Monit. Assess*., In press, ISSN: 0167-6369.

Nelson, E.J. (2006). CE-QUAL-W2 Interface, In: *WMS v8.0 Tutorials*. Accessed on February 23th, 2011, Available from: http://www.cequalw2wiki.com/cequalw2/images/9/99/WMS_8.0_CE-QUAL-W2_Tutorial.pdf

Obregon, O., Chilton, R.E., Williams, G.P., Nelson, E.J., Miller, J.B. (2011). Assessing Climate Change Effects in Tropical and Temperate Reservoirs by Modeling Water Quality Scenarios, *Proceedings of the 2011 World Environmental and Water Resources Congress*, paper 407, Palm Springs, USA, ISBN: 978-0-7844-1173-5.

Rangel-Peraza, J.G., de Anda, J., González-Farías, F.A. & Erickson, D.E. (2009). Statistical assessment of water quality seasonality in large tropical reservoirs, *Lakes & Reservoirs: Research and Management*, 14(4), 315-323, ISSN: 1440-1770.

Torres-Orozco, R.E. (2007). Batimetria y morfometria, In: *Limnologia de las presas mexicanas. Aspectos teoricos y practicos*, J.L. Arredondo-Figueroa, G. Diaz-Zabaleta and J.T. Ponce- Palafox (Eds), pp. 3-19, AGT Editor S.A., Mexico, D.F. ISBN: 9789684631366.

Wetzel, R.G. (1993) Limnologia. Fundacao Calouste Gulbenkian. Lisboa, Portugal, 919 p. ISBN: 9789723106046.

Wetzel, R.G., Linkens, G.E. (2000). Limnological analyses. Springer. NewYork, USA, 429 p. ISBN. 0387989285.

Xu, X., Tan, Y., Yang, G., Li, H., Su, W. (2011) Soil erosion in the Three Gorges Reservoir area. *Soil Research*. 49, 212–222. ISSN: 1838-6768.

Zhao, X., Shen, Z.Y., Xiong, M., Qi, J. (2011) Key uncertainty sources analysis of water quality model using the first order error method. *Int. J. Environ. Sci. Tech*. 8 (1), 137-148. ISSN: 1735-1472.

6

Palynology as a Tool in Bathymetry

Cynthia Fernandes Pinto da Luz
Instituto de Botânica, Núcleo de Pesquisa em Palinologia, São Paulo
Brazil

1. Introduction

1.1 Rationale

Palynology is the science that studies the ontogeny, structure, dispersal mechanisms, deposition and preservation of spermatophyte pollen grains and spores of fern, mosses and liverworts in different environments (Erdtman, 1943). In a wider sense it involves the study of palynomorphs, a term created by Tschudy & Scott (1969) to define the organisms resistant to drastic chemical treatments used in Palinology as, e.g., some microscopic algae. This is because the composition of the palynomorphs external wall, constituted by sporepollenin, an organic polymer composed of carbon, oxygen and hydrogen, probably is the most resistant organic matter of all living beings, and has remained unaltered for millions of years, even after the death of the cell content. Due to its chemical and microbiological degradation resistance, palynomorphs have the potential to become microfossils in sediments (von Post 1916, Zetzsche 1932, Chaloner 1976, Brooks & Shaw 1978, Moore *et al.* 1991, Takahashi 1995).

The palynological analysis of the sediments is essentially based on plants reproductive strategy of abundantly release pollen grains during the flowering season, and spores during the sporophytic phase, in some cases in billions per m^2. Apart from their different functions in plants, pollen and spores can be used in the reconstruction of recent and past vegetation, as they are easily carried by the wind due their minute size (ranging from a hundredth to a tenth of a millimeter in diameter) and be transported to high altitudes by vertical currents, remaining in the atmosphere for days, weeks or even months moving long distances to precipitate as "pollen rain" over land and water. After falling from the air to the soil or water, a number of factors affect their conversion into microfossils, before and during their sedimentation. Pollen grains and fern spores never accumulate in their original form when deposited in the sediments. This includes factors that can destroy spores and pollens in sedimentary deposits. Aware of these facts, researchers have developed several studies in order to observe the spore-pollen deposition in various current environments to serve as the basis for paleoecological studies.

As previously stated, current and past vegetation records can be preserved only where pollen and spores have accumulated as microfossils through time. As oxygen is the main destructor of organic matter, the deposition environment has to be free of this element (or present in small concentrations only) in order for pollen grains and spores be preserved after sedimentation, i.e. an anaerobic environment such as the subaquatic one. They are integrated in the water as silt and clay size particles and get exposed to the same laws of particle movement in fluids, subject to water circulation dynamics that function as transport agent before their sedimentation in rivers, deltas, estuaries, lakes, bays, lagoons and open seas.

The accumulation of pollen and spores of recent and fossil deposits in sediments of aquatic ecosystems is a rich source of ecological information as the climatic evidence is indirectly contained in the biological data. The vegetational changes in the fossil record may have been caused by climate change. However, one should keep in mind that not all changes of pollen, spore and algae accumulation are necessarily caused by climatological factors. Sedimentary records can incorporate other kinds of evidence that can interfere in the palynological analysis of the vegetation changes. These interferences can mask the climatological data such as those caused by human activity in the vegetation or fire and insect infestations that require the analysis of other indicators, such as coal, which should be added to pollen spectra of anthropocoric plants.

The inherent characteristics of each palynomorph also affect the accumulation in subaquatic deposits, both in space and time, in number and quality of the sedimented material, causing under- or over-representation of specific types, depending on the sample local. Changes of the depositional processes that result in alterations in the accumulation and preservation of the palynomorphs and changes on the water level affecting the local flora succession, among others, are examples of interferences in pollen and spore frequency that can challenge the interpretation of regional flora and climate by means of Palynology. After sedimentation, the resuspension of previously deposited palynomorphs, and the convergence of these to other parts of the drainage basin due to currents and winds, also cause alterations in their frequencies, both in the central area and margins. In addition, the preservation of sedimentary deposits differs with each palynomorph type, which can result in its destruction before or during its incorporation in the sediment (Stommel 1949, Cushing 1966, Davis 1968, Davis *et al.* 1971, Davis & Brubaker 1973, Peck 1973, Bonny 1980, Delcourt & Delcourt 1980, Davis *et al.* 1984, Campbell 1991, Moore *et al.* 1991, Campbell 1999).

It is evident that defining the vegetation homogeneity or heterogeneity using pollen grains and spores is not a simple matter. As there is no constant between the release rate and the rate of spore and pollen grain accumulation it is not possible to make a direct correlation with the productivity of each parent plant. Even in underwater environments with stable sedimentation the inference regarding an aspect of certain vegetation based on palynological representation is inconsistent if the different processes that cause the possible spatial variations in the deposition of pollen, spores and algae aren't taken in consideration. Here bathymetry has a fundamental role. Any change in the frequencies and concentrations of pollen and spores may indicate changes in bathymetry and in the water volume.

This work aims to show the importance of an accurate spatial analysis of the deposition rate of pollen and spores in subaquatic sediments for a precise correlation with the source vegetation. In the preface is provided a brief description of the characteristics of pollen and spores, their different release mechanisms from the parent plants and possible factors that affect their conversion into microfossils. Details of the methods used for extracting palynological information in surface sediments and the criteria for data selection are defined in the next section. Finally, the results obtained from palynological analysis of surface sediments in some locals from the coast of Rio de Janeiro are compared and classified into groups that represent the major influences in sedimentation of pollen and spores, i.e., the bathymetry, winds and currents.

1.2 Primary Differential Processes – production and release mechanisms of pollen grains and spores of fern, mosses, liverworts and algae

The "Primary Differential Processes" are composed of different ways of production and release mechanisms of pollen grains, spores and algae that are peculiar to each parent plant added to the initial depositional influences unique to each sedimentary environment.

1.2.1 The pollen grains

Pollen is the microgametophyte of Gymnosperms and Angiosperms. It is the fecundate element, the cell that contains the male reproductive nucleus. It is largely produced in flower anthers of monocots and dicots, and in male cones of Gymnosperms, both constituting the pollen grain contents of the pollen sack (= anther locules). The pollen grain is an essential element of the sexual reproduction of plants and it needs to reach the female part of the flower or cone to germinate and form the pollen tube that takes the male nucleus to the ovule (megagametophyte). The fusion of pollen and ovule nuclei originates the embryo and its involucres, constituting the seed which is the disperser agent of the species.

Until de 17th century nothing was known about pollen and its role as a fecundate source. It was only in the 18th century that the first successful observations and experiments had begun and proved that fruit development did not happen without pollen (Wodehouse 1935). In the 19th century, the microscope equipment was 500 times more efficient than magnifying glasses and made possible the visualization of the pollen external-wall (exine) which led to the discovery that it was ornamented and had apertures and other morphological characters, frequently similar within a species, allowing the identification of the plant that had produced it (Salgado-Labouriau 2007).

Pollen grains are produced after meiosis, when each pollen mother cell divides into four haploid cells. These frequently split and each pollen grain remains isolated from the other. However, pollen grains of some species remain attached in groups of two, four, more than four and in pollinium (Fig. 1). Pollen grains are involved by exine, a sporopollenin external-wall, composed by the sexine and nexine layers. Due to its resistance to chemical and microbiological attacks, pressure, and temperature changes, it is preserved for millions of years. In sedimentary rocks Gymnosperm pollen from the Paleozoic Era, more than 300 million years ago, can be found and Angiosperm pollen from the Upper Cretaceous, more than 100 million years ago (Traverse 1988).

The morphological analysis of pollen grains involves a series of descriptions (Barth & Melhem 1988, Punt *et al* 2007). The literature is rich, including catalogs and treatises on pollen morphology (Erdtman 1952, 1957, 1965, 1971, Faegri & Iversen 1950, Melhem *et al.* 1981, Roubik & Moreno 1991, Salgado-Labouriau 1973, among others). With this knowledge at hand it was possible to verify that unrelated plants can have similar pollen types and some plant families may have more than one morphological pollen type. There are monomorphic species, i.e., species with a basic morphological pollen type other that are dimorphic, with two types, or polymorphic with several. That's why the identification of pollen grains remains in the "Pollen Type" category in microfossil studies. The pollen type represents an artificial grouping based on pollen morphological characteristics within or between families. Frequently, similar species, varieties and subspecies within a species have the same pollen type (Salgado-Labouriau 1973).

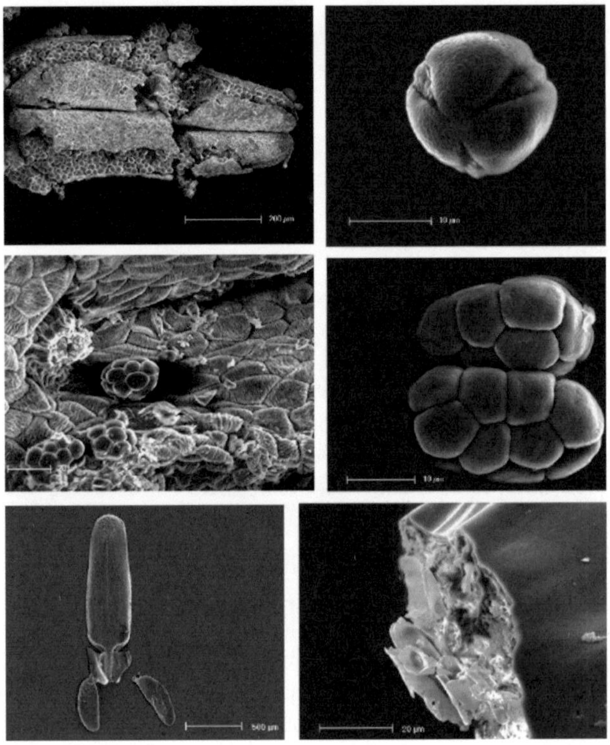

Fig. 1. Scanning electron microscope images of anthers, pollen grains and pollinarium.

Top: anther with pollen grains of *Zornia diphylla* (L.) Pers. A single pollen grain of *Z. diphylla*. Middle: interior of the anther of *Stryphnodendron adstringens* (Mart.) Coville with polyads. Two polyads of *S. adstringens*. Bottom: general view of the pollinarium of *Oxypetalum insigne* (Decne.) Malme. Detail of the pollen grains inside the pollinium of *Oxypetalum capitatum* subsp. *capitatum* Mart.

There are different mechanisms by which plants release their pollen grains through biotic and abiotic agents in a wide range of specializations in order to avoid genetic losses due to environmental interference. The transport of pollen grains from the anther to the stigma is called pollination, and it can happen directly or not. In Angiosperms there are different pollination mechanisms. Direct pollination occurs in autogamous plants (pollination of the same flower) where biotic or abiotic agents can help in pollination, but they are not essential since pollen is received on the stigmatic surface of the same flower. This is the case of cleistogamous plants, where flowers are still closed when pollination happens. The pollen of cleistogamous plants is rarely seen in sedimentary records.

Indirect pollination occurs in allogamous plants (pollination between different flowers) where biotic and abiotic agents play an essential role. Abiotic agents can be water and wind, while biotic agents can be different groups of animals (Fig. 2). In hydrophilic plants pollen is taken to the stigma of another flower by water. A mass production of pollen is necessary, however, pollen of this type of plant does not frequently show any sexine or nexine, thus not preserved in sediments.

Zoophilous plants use animals as pollinators. For the efficiency of this type of pollination the production of pollen of these plants is generally reduced. Zoophilous and ambophilous taxa (plants pollinated by both wind and animals) frequently occur in the pollen assemblage of sedimentary rocks, but they are consistently underrepresented (absence of pollen in the sediment while the parent plant does exist in the vegetation or by the abundance of pollen is much smaller in the sediment than the abundance of the parent plant in the vegetation). If no abiotic agent interferes, the pollen concentration of zoophilous taxa can be high in the ground next to the plant, and its presence in the fossil sediment can indicate proximity of the parent plant.

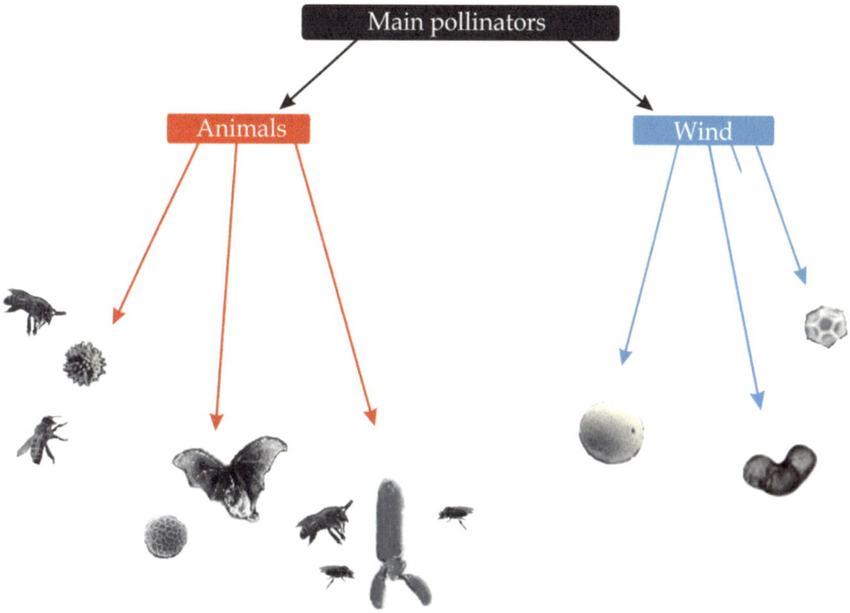

Fig. 2. Zoophilous pollen grains generally show complex sexine structures and use oils to stick to the body of the animal and may be a single grain or grouped into tetrad, polyad and pollinarium. The typical pollen grain pollinated by the wind has an aerodynamic shape with simple structures of sexine ornamentation and sometimes hollow spaces.

Anemophilous plants use wind to disperse pollen grains and include all the Gymonsperms and a substantial number of Angiosperms. With very few exceptions most palynomorphs accumulated in Quaternary lacustrine sediments consist of terrestrial plants pollinated by wind (Jackson 1994). Anemophylous pollen dominates even in tropical regions where the trees pollinated by insects are more abundant in the forest canopy (Kershaw & Hyland 1975, Colinvaux *et al.* 1988). As a large part of the anemophilous pollen don´t fulfill their role in pollination, they are deposited in large amounts everywhere meanwhile zoophilous pollen enters the environment attached to animals that generally avoids landing in certain areas (for example, in water surfaces). Finally, anemophylous pollen is aerodynamic, light, with sizes between 5-100 µm, and they are better suited for transport by the wind longer distances than zoophilous pollen, frequently big and dense, that tend to stick together with other pollen grains (Whitehead 1969, 1983). This way, even if zoophilous pollen escapes from flowers to the air they will not travel long distances until deposition in the sediments.

1.2.2 The spores

Spores are the disperser cells of Cryptogams (mosses, liverworts, ferns, etc) which contain the genome and that asexually develops into a new gametophyte. In ferns (prothallus) the gametophyten is reduced, normally subterranean, and self-sufficient, while in mosses and liverworts the gametophyte represents the most developed generation, which we know as the "plant".

The gametophytes of mosses and liverworts' produce antherozoids and eggs that give origin to the sporophyte once fecundated, which grow from its own archegonium (like the ovary in plants) in the shape of a capsule - the sporangium - at the edge of a peduncle. The sporophyte attached to the parent plant produces haploid spores that gives origin to new plants after dispersion. In ferns, on the other hand, the sporophyte is independent, which is what we know as the "plant". The spores are produced in structures on the abaxial side of the leaf, or frond, called sorus, where the sporangia are found.

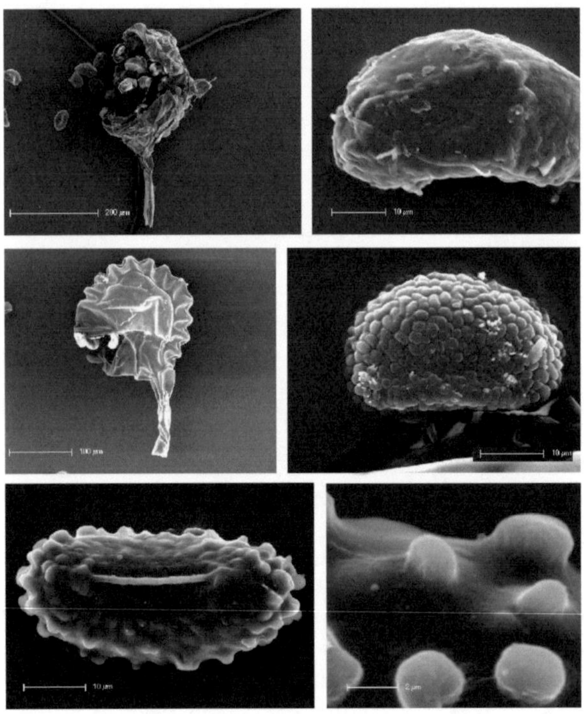

Fig. 3. Scanning electron microscope images of fern spores and sporangia. Top: sporangium of *Serpocaulon glandulosissimum* (Brade) Labiak & J. Prado. Monolete spore of *S. glandulosissimum* covered with the perine layer. Middle: sporangium of *Serpocaulon sehnemii* (Pic. Serm.) Labiak & J. Prado. Monolete spore of *S. sehnemii*. Bottom: detail of the monolete spore laesura of *Serpocaulon richardii* (Klotzsch) A. R. Sm. Detail of the verrucae ornamentation of the monolete spore of *S. richardii*. (Photos provided by Dr. Carolina Brandão Coelho and Dr. Luciano Mauricio Esteves, Instituto de Botânica, Brazil)

The spore, just like the pollen, has a resistant external wall composed by sporopollenin. Fern spores are found in sedimentary rocks from mid Silurian, more than 400 million years ago (Traverse 1988). The palynological analysis of Cryptogam spores is very similar to that from pollen grains, but there is less morphological variance and specific nomenclature (Barth 2004, Erdtman & Sorsa 1971, Tryon & Lugardon 1990, Lellinger 2002). The spore is usually a spherical, tetrahedral or reniform structure, frequently with elaborated ornamentation patterns (Fig. 3). Mature spores are always in isolated grains, and can have leasures on the proximal face. These leasures are important for its identification. Monolete spores have only one leasure, trilete spores show a leasure in a Y shape, and aletes do not have leasures and are commonly found in mosses (Cruz 2004).

In the mosses and liverworts where the sporophyte production is high, spores are produced and released over many months (e.g. *Anthoceros*). Generally in terrestrial plants, spores are dispersed by the wind and capable of resisting long period of drought. However, it is only when they fall over humid and suitable surfaces that they will absorb water and germinates.

According to Tuomisto & Poulsen 1994 (*apud* Graçano *et al.* 1998) edaphic specializations in some fern species justify its use as soil fertility indicators, where patterns of geographical distribution should be considered in ecological studies.

1.2.3 Chlorococcales algae

There are relatively few algae *taxa* with spores, cysts or other resistant forms that are preserved in sediments, that stands out in the geological history, from 500 million years ago (Brenner & Foster 1994, Jansonius & Mcgregor 1996). Algae are relatively recently used in paleoecological interpretations (van Geel 1976, van Geel & van Der Hammen 1978, Salgado-Labouriau & Schubert 1977, Luz *et al.* 2002). Algae from the Chlorococcalles order are the most abundant microfossils found in lake and swamp sediments, due their resistant external wall of sporopollenin. It is an order of green algae that includes both unicellular and colonial species (Fig. 4).

All Chlorococcales have an endogenous asexual reproduction (vegetative), where the number of daughter cells or colonial cells is determined by the number of cleavages of the mother cell. Some have solitary cells isolated inside the colony (Botryococcaceae for example), while others have directly united cells that form a *coenobium* (Coelastraceae, Hydrodictyaceae and Scenedesmaceae, for example).

The life period of unicellular algae is probably measured in hours or days. Asexual reproduction in algae usually occurs under favorable conditions, while sexual reproduction occurs when conditions are less favorable. If the algae doesn´t reproduce the protoplast dies and the remaining cell wall sinks and deposits in the sediment. In some species there is also a type of vegetative reproduction where somatic cells can change by adding a thick wall, and these cells can function as resistance spores (cysts), dormant during hostile periods, while all other somatic cells die. The dimensions of individual cells and the *coenobium* depend on the growth rate which in turn depends on environmental factors (Brenner & Foster 1994, Jansonius & Mcgregor 1996).

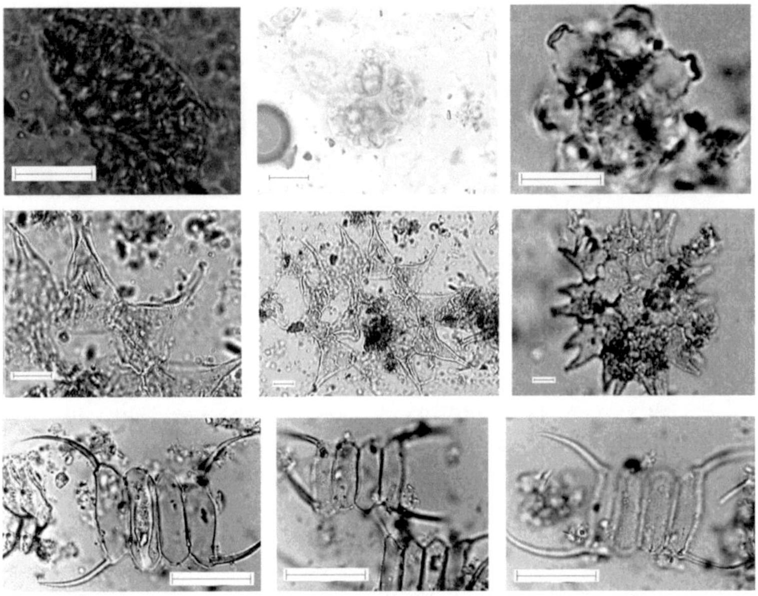

Fig. 4. Light photomicrographs of Chlorococcales algae. Top: colonies of *Botryococcus* sp1, *Botryococcus* sp2 and *Coelastrum proboscideum* Bohlin. Middle: detail of the coenocytes of *Pediastrum* sp, *coenobium* of *Pediastrum* sp and *Pediastrum duplex* var. subgranulatum. Bottom: colonies of *Scenedesmus protuberans*, *S. magnus* and *S. ohauensis*. Scale = 10 µm.

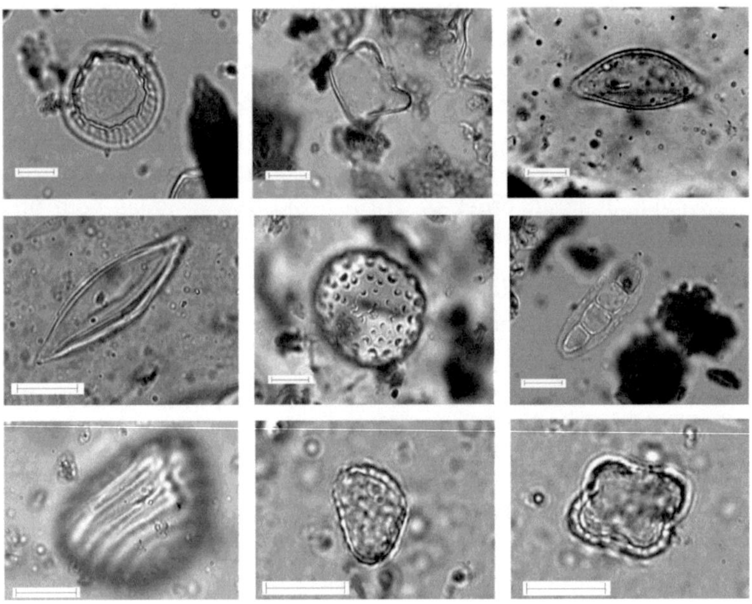

Fig. 5. Light photomicrographs of Zygnematales and others algae. Top: zygospores of *Debarya* (De Bary) Wittrock, *Mougeotia* C. A. Agardh and *Spirogyra* Link sp1. Middle: zygospores of *Spirogyra* Link sp2, *Zygnema* C. A. Agardh and a not identified algae. Bottom: Other forms of algae. Grain of *Incertae sedis* (*Pseudoschizaea rubina* Rossignol ex Christopher) and two forms of *Tetraedriella jovetti* (Bourelly) Bourelly (Xanthophyceae). Scale: 10 µm.

1.2.4 Zygnematales algae

Just like the Chlorococcales, many zygospores from the Zygnemataceae family have a resistant external wall, composed of sporopollenin (Fig. 5). This family unites the algae formed by cylindrical cells associated to filaments, floating and free-living, although *Mougeotia, Spirogyra* and *Zygnema* may also grow adhered to substrates (Smith 1955, Kadlubowska 1984 *apud* Dias 1997). They are inhabitants of shallow waters of freshwater lakes and ponds. However, they have also been found in brackish and saline environments. They easily grow in lentic environments (shallow and stagnant waters or in higher waterbodies, permanent or temporally), lotic, phytotelmic, terrestrial bromeliculous and subaerials (wet soil and peat) (Dias 1997).

1.2.5 Vegetational representation in surface sediments

Knowing that the assemblage of pollen and spores recovered from the sedimentary record cannot be directly interpreted as an accurate reflection of parental vegetation, researchers recognized the need to know in detail the source area of pollen and spores (the vegetation area from which most pollen and spores derives) to interpret the patterns observed in the palynological diagram curves. The concept of source area originated from considerations of von Post (1967) from a transect in Sweden regarding location and spacing of the depositional site in relation to long-distance pollen transport. Differences in the efficiency of the dispersion of pollen grains and spores mean that many of them found in the depositional site may have originated in plants located in a wide geographical area, transported by winds or rivers. Therefore, the percentage analysis of the pollen grains record requires knowledge of the abundance of plants found at the depositional site, both locally and regionally, helping to better understand issues regarding overrepresentation (abundance of pollen in the sediment is much higher than the abundance of the plant at vicinity) and underrepresentation of certain spore and pollen types. The abundance of the parent plant in the landscape can be perceived and described in a variety of ways. The ecologist expresses abundance of plants per unit area in terms of stem density, total biomass, coverage area projected vertically, etc. The intensity of pollen and spores of the source area is another way of describing the abundance of the plant. Intensity of spore-pollen in the source area depends on many factors, however, in many cases, it is inversely measured by the distance of the depositional site to the vegetation where the *taxon* in question is located. Several authors have developed studies on the source area (Tauber 1965, Janssen 1966, Andersen 1970, Janssen 1973, Tauber 1977, Bradshaw 1981, Jacobson & Bradshaw 1981, Parsons & Prentice 1981, Prentice 1985, Prentice 1988, Calcote & Davis 1989, Jackson 1990, Jackson 1991). Since these studies used "pollen and spores traps" to relate to the sedimentary basins, which offer a "sedimentary environment" very different from reality, quantification is limited. Comparative studies of pollen deposited today and existing vegetation in the vicinity highlighted the importance of certain choices of parameters for the calibration of the relationship pollen/vegetation and for the interpretation of spore-pollen records as a whole (Webb *et al.* 1978, Parsons *et al.* 1980, Bradshaw & Webb 1985, Prentice *et al.* 1987, Jackson 1990). The selection of the size of the depositional site and the choice of Spore-pollen Sum (i.e. which *taxon* to be included in the diagrams) are very important for the analysis of the sedimentary record of the vegetation. The variations between the spore-pollen assemblages can, in many cases, be related to vegetation patterns in the spatial scale from 10^2 to 10^3m, however, the size of the drainage basin affects the pollen representation of this vegetation. The results of simulations on the vegetation spatial scale represented by pollen grains and spores deposited in lake sediments indicated that this scale may seem homogenous in palynological records even when its actual pattern is heterogeneous and uneven. This will depend on the size of the depositional site in relation to the size of the vegetation patches existing in the surroundings. The larger the lake, the more the pollen assemblage will be influenced by the extra-local and regional components (Janssen 1966, 1984; Andersen 1970; Calcote & Davis 1989; Jackson 1994). The data obtained corroborate the empirical knowledge that smaller lakes are especially suitable for the reconstruction of local vegetation, while larger lakes are more appropriate for the reconstruction of regional vegetation and climate.

Sugita (1993) points put the importance of recognizing the differences between the areas of pollen source in the central part of the depositional site and the pollen deposited over the entire surface. The author constructed a differential deposition model based on the model of pollen transport to depositional basins suggested by Prentice (1985, 1988). Several assumptions and mathematical equations suggest that the local distribution of plants around the lake has great influence on the pollen assemblage found throughout the lake basin, and that pollen deposition decreases slowly from the margin to the center of the basin, and the center tends to present a lower deposition, since it is farther from the closer area of pollen source. In this model, the mean of pollen input of the entire basin would involve the effect of higher accumulation of pollen near the margins. However, for the author, even in locations without significant pollen input from aqueous streams and tributaries entering the basin, the re-suspensions and redirection of the sediment to the deepest part can generate high rates of pollen deposition in the deeper areas increasing differences in pollen deposition in the entire basin.

Regarding local components, it is assumed that much of the pollen deposited in aquatic plants derives from plants that grow in the lake. As pollen and spores of local plants are usually overrepresented in the depositional site, any change in their frequencies and concentrations may indicate changes in bathymetry and in the water volume. Still, the representation of aquatic plants in spore-pollen assemblage is highly variable, depending on its abundance, lake extent as well as if the sample was obtained from benthic or coastal sediments (Janssen 1966). The pollen analysis of coastal sediments can indicate changes in aquatic vegetation resulting from disturbances in the ecological succession and in the water level, because their percentage in the assemblage is reduced when horizontal dimension of the water body decreases (Janssen 1967, Birks *et al.* 1976, Jackson *et al.* 1988). According to Jackson (1994), many submerged and floating aquatic plants are pollinated by insects or have their pollen dispersed by water. Although many important *taxa* are anemophilous (Cook 1988), the pollen production of these plants is low and dispersal is often limited. In other emerging aquatic plants, pollination is anemophilous and they are high-pollen producers (e.g., *Typha*, Poaceae and Cyperaceae) and, thus, high amounts of pollen are observed where these plants grow. The challenge is to determine how much pollen of these *taxa* represents local aquatic plants or how much originates from regional terrestrial plants.

The phytoplanktonic community in watersheds is conditioned to dynamic processes related to the physical and chemical instability of their waterbodies, among which salinity fluctuations and variations in the concentration of nutrients stand out. These factors will regulate populations and interfere with the phytoplankton succession. These fluctuations are related to the water circulation which is a reflex of the hydrography and the concentration of the suspended material, besides being conditioned to the annual cycle of entry and exit of water in the system (evaporation relation/precipitation). The succession of Algae usually begins with green algae, and then by blue algae in eutrophication processes. In large scale, fluctuations in phytoplankton populations in coastal lakes and lagoons were linked to variations in the sea level from its origins to the present. If in any moment there was a water entry with a higher salinity than the lacustrine environment, it would have caused a total change of the richness, diversity and density of certain algae genera. The characteristics of phytoplankton in these environments, when under direct marine influence, show high values of biomass, a high productivity and low diversity (Margalef 1969(Comin 1984, Margalef 1969 and Odebrecht 1988, *apud* Huszar & Silva 1992).

1.3 Secondary Differential Processes – differential sedimentation, preservation and reworked palynomorphs in rivers, lakes, estuaries and deltas

The "Secondary Differential Processes" are the various depositional influences after sedimentation of pollen and spores, each one unique to a specific sedimentary environment.

The interest in palynological sequences that record the dynamics of vegetation in the Quaternary period has led palynologists to seek depositional environments of good preservation for pollen and spores with stable, continuous and datable deposition. Overall, it is assumed that in these environments, variations of palynological records attributed to depositional processes are small compared with changes in the abundance of plant species and in the production and release of pollen and spores. However, even in environments with stable sedimentation, palynological studies show that in the horizontal gradient (transect) the deposition of palynomorphs display different patterns of location. The spatial differences in the abundance of sedimented pollen grains, fern spores and algae are striking, even in lakes without tributaries and located nearby each other. Moreover, in certain basins, depositional processes are the main causes of interference in the palynological records. In aquatic environments, depositional processes occur because spatial variations in rates of fern spores and pollen grains accumulation over time ("influx" = amount of pollen that falls each year in cm^2 of soil) are influenced, among other factors, by the seasonal differences in the "input" of pollen and spores and the hydraulic selectivity ("sorting") the various types of palynomorphs suspended in water, according to their different morphologies and densities, which cause differences in the sinking speed. In addition, there are physical and chemical characteristics inherent to each aquatic environment such as the intensity of water currents and vertical movements of water (seasonal and daily) caused by variations in temperature and density. Types of sediments deposited at the bottom and their movements to another part of the basin as well as the bathymetry and the intensity/direction of dominant winds that can cause resuspension of previously deposited material, also influence the pollen and spores sedimentation. Thus, the hydrodynamic distribution of pollen grains and fern spores, as well as other particles, produces differences in the quality and in the total of accumulation rates of these particles in different locations of the basin. Therefore, the spatial analysis of fern spores, pollen grains and algae deposition in the surface sediments of the bottom of water bodies show different patterns from place to place and can help in sedimentological studies (Davis 1968, Davis *et al.* 1971, Peck 1973, Lehman 1975, Bonny 1980, Davis *et al.* 1984, Moore *et al.* 1991, Sugita 1993).

Jackson (1994) highlighted the care to be taken in interpreting the temporal changes observed in the recent and fossil spore-pollen assemblages in sediments, because to directly correspond to the respective changes in the intensity of the spore-pollen *taxon* involved (number of pollen grains or fern spores produced by a *taxon*, per unit of land area, per unit time) all other aspects should remain constant (e.g. spore-pollen dispersion and spore-pollen deposition), which never occurs.

Stommel (1949) studied the behavior of particles affected by the action of water currents produced by winds and demonstrated that the number of long-axis vertices in parallel formed at the water surface and the direction of dominant winds cause different types of distribution of these particles. This distribution occurs according to the sinking speeds of each, and the particles can only sink if the sinking speed is close to the maximum upward speed of the water. Potter (1967) found that even with the influence of dominant winds in a single direction, the pollen grains deposition in the sediment of the bottom of water reservoirs varies among the *taxa*, at the banks as well as in the center of the reservoir. Contrary to the expectation that dominant winds could cause higher pollen concentration at the opposite margin, results show a complex differentiation of pollen accumulation in bottom sediments.

Hopkins (1950) investigated the spatial differential sedimentation in lakes regarding the sinking speed of certain pollen grains and found that the *Pinus* pollen sinks less quickly than the *Quercus* pollen suggesting that this is due to differences in the size of pollen grains. Bradley (1965) pointed out that the sinking speed of many particles in a lake, such as the diatoms, is too slow to allow them to reach the bottom of deep lakes in the same year that they were formed and are, thus, subject to various transport mechanisms. Davis (1968) observed that the pollen types with low sinking speed (smaller sizes) are preferentially deposited in shallow areas of the coast of a lake. The author also showed that in the bottom sediment, more than 80% of the pollen was derived from sediments that had previously been deposited elsewhere in the lake (redeposition). Davis & Brubaker (1973) based on the Stommel theory to affirm that water circulation in a lake affects pollen sedimentation according to the different morphological types, and even small differences in sinking speed of each pollen type would cause great effects on the distribution of these in aquatic environments and, therefore, the total accumulated pollen grains are different, for example, in several parts of a lake.

Merilainen (1969) suggested that the epilimnion current (top-most warmer and less dense layer in a thermally stratified lake) affect the deposition patterns of diatoms in sediments.

Davis *et al.* (1984) demonstrated that there is a tendency for deeper areas of a lake to accumulate sediments faster than the shallower ones, thus resulting in large variations in accumulation rates of pollen among samples. Faegri *et al.* (1989) reported that the differential spatial distribution is due to bathymetric differences and the effective capacity of pollen grains and spores to accumulate in accordance to their sinking speeds. Sugita (1993) suggested that the pollen "input" proportion of a total area in a lake basin is higher than the pollen deposition at the center. The author also noted that the diameter radius of the spore-pollen deposition is often longer in lighter pollen grains and spores than in the heavier ones, which may have more pronounced differences in the deposition percentages in relation to the total area of a lake basin.

Pollen reworking is generally an indicator of an unstable environment, and instabilities can always occur on a time scale. Several authors (Davis 1968; Davis *et al.* 1971; Peck 1973; Bonny 1980; Davis *et al.* 1984), concerned with palynological analyses of lake cores based on estimates of pollen deposition from a single central point of the sedimentary basin, highlighted that the internal processes of a lake redistribute the pollen originally deposited on the bottom surface. These may causes mixtures through the resuspension of previously deposited pollen and the pollen existing in the water prior sedimentation through the "sediment focusing" to the deeper area of the basin. These processes of secondary importance generate the differential deposition and an assembly of palynomorphs in which each type has a differential preservation (Campbell 1999). Dupont (1985) suggested that a differential removal of pollen in aquatic environments would be solely due to the water movement as surrounding areas have different incidences of pollen grains and spores, as in the case of a canal bed and its margin. However, Campbell (1999) explained in detail the reworking process of pollen grains and spores, and the removal of the oldest ones from the deposit would operate in four ways:

- a complete removal of the deposit and total redeposition.
- a partial removal and total deposition of the reworked fraction.
- a complete removal of the deposit and partial deposition (in this case, a fraction of pollen grains and spores could be destroyed by transport as suggested by Fall (1987) in the case of a fluvial transport).
- a partial removal and partial deposition.

The author suggested four fundamental processes that could occur in the differential reworking:

- differential resuspension of the original deposit.
- differential transport.
- differential capture in the receiving area.
- differential preservation during transport.

As an example, the passage of water or wind over a surface could resuspend the pollen grains and spores in a different way, leaving the heavier ones in relation to hydrodynamic (or aerodynamic) behind and moving the lighter ones to a new deposit. This type of resuspended pollen assemblage presumably occurs constantly in the environment. In cases where redeposited palynomorphs are not obviously older than those of the original deposit, their presence becomes very difficult to detect.

For Chmura *et al.* (1999) the fluvial transport of palynomorphs provides a more inclusive vision of the vegetation than the aerial transport alone, corresponding to a palynological assemblage of the vegetation found in the drainage basin, including high amounts of pollen and spores that would not be readily available to the anemophilous transport (such as herbaceous plants). The author noted that, on the contrary, deposits of pollen and spores in lakes without tributaries and located far from estuaries are much more influenced by anemophilous plants, particularly wind-pollinated tree species. The author suggested that in the fluvial transport, the deposits of banks, sand and bars existing in the way, can also introduce spores and pollen of local plants (local component) to the pollen assemblage, as well as lead to the resuspension of these deposits. Obviously, this supply of pollen and spores directly from rivers is dependent on its geomorphology. A sinuous river will be much more exposed and receive more pollen and spores from the banks than a straight course river.

Muller (1959), Cross *et al.* (1966), Traverse & Ginsburg (1966), Heusser & Balsam (1977), Heusser (1978) and Traverse (1988) when studying the ocean transport of pollen and spores found that they are transported by currents and deposited together with silt preferably at the river mouths. The contribution of plants growing on the river banks dominates the spore-pollen assemblage, since most of it is introduced into the ocean by fluvial transport. In the absence of ocean currents, abundance of pollen and spores decreases with the distance from the coast, but the turbulence of the sea water and waves are important factors in the redistribution of palynomorphs. Those that settle far from the coast are those able to float longer, such as the bisaccate pollen grain of conifers that have hollow cavities formed by an expansion of the exine. But Wang *et al.* (1982) observed that the surface sediments of the Yangtze River mouth (China) showed a low concentration of palynomorphs, while high concentrations were found far from the coast. Traverse (1988) suggested that this pattern could be explained by the local hydraulic turbulence, since the pollen deposition at river mouths is primarily controlled by currents and hydraulic sorting according to the sizes of pollen grains and, therefore, the ordering of the pollen assemblage should follow a distance gradient from the delta (Fig. 6).

Another point that deserves attention is the pollen and spores destruction before or even during sedimentation. Pollen grains and spores are subject to various weathering and decay processes, from the time of anther dehiscence and sporangia until the deposition time (Campbell 1999). During periods of soil erosion, pollen and spores can become incorporated into river and lake sediments. As a result, the contemporary vegetation may be poorly represented by the palynomorphs in the sediments because of the reworked component of in washed pollen and spores. This assemblage of reworked palynomorphs generally presents itself with several levels of exine deterioration. The differential preservation is often recognized by the tendency that the assemblage of pollen grains and spores shows in terms of their poorly preserved condition and abundance. Analysis of the level and type of deterioration is very important in assessing the sedimentation conditions to which pollen and spores have been exposed because changes in taphonomic process can influence the composition of the palynological assemblage, producing variations independent from changes in vegetation (Pennington 1996). Differences in the preservation of pollen grains and spores during transport have been reported in many studies, particularly those related to damage caused by collisions of the pollen grains and spores in fluvial transport (Catto 1985, Fall 1987). However, Campbell (1991) showed that this type of damage is minimal, because the greater damages occurred during fluvial transport are likely to be originated from oxidation and dryness in temporal deposition areas along the time.

It is known that the endexine is more resistant to oxidation than the exine (Rowley 2001). The type of deposit, to which pollen grains and spores are eventually incorporated, will affect the assemblage of palynomorphs, meaning that the differential degradation may continue after the deposition. The oxidative-reduction potential (Eh) of the depositional environment is affected when the sediments with low Eh are more favorable for the pollen preservation. The adverse effect of soil pH on the pollen and spores preservation occurs in soils with pH above 6.0 (alkaline) where the pollen is not usually preserved and when preserved the identification to the taxonomic level becomes impossible due to the poor

Fig. 6. General representation of differential deposition of palynomorphs in a river mouth: large size grains are deposited in the delta, near to the river mouth; medium size grains are deposited at sites of transitional depths of the delta and the smallest ones or those with hollow cavities such as bisaccate pollen grains of *Pinus* and *Podocarpus* continue to float longer, placing it farther away with the currents or, in the case of shallow lakes, on the opposite side of the dominant direction of the wind.

visualization of the exine (Pennington 1996). In the destruction process of pollen grains and spores, the biochemical attack of bacteria and fungi also plays a very important role (Elsik 1971). The speed and damage extent caused by all these factors are in many cases related to the genetics of pollen and spores such as the low amount of sporopollenin of the exine that generates greater instability in their preservation in the sediment (Havinga 1964). The sculpturing elements of the exine may also provide greater or lesser resistance to the attacks, as in psilate pollen grains (usually with thin exine) that are less resistant to oxidation.

Moore *et al.* (1991) noted the importance of observing the level and type of pollen deterioration of any palynological sample.

They summarized the work of Cushing (1967) and Delcourt & Delcourt (1980) on four deterioration types of pollen grains and spores (Fig. 7):

- **Corrosion** is characterized by the completely perforated exine, as a network of circular holes or all the tectum (layer of the sexine) is removed, leaving an exposed surface of scabrate appearance. Sometimes only the outermost layer of the sexine is affected and may be slightly excavated. This type of deterioration is more intense in peat bogs deposits. The most usual cause for it is the microbial activity. Provided that the growth of bacteria and fungi activity occurs in conditions of, at least, periodical aeration, the implication of such deterioration in aquatic environments is that before or after the deposition the pollen grains and spores were exposed to oxygen. The microbial attack, particularly by anaerobic bacteria, can remain in humid and flooded sediments, but with a lower rate.

- A general reduction in exine thickness characterizes **degradation**. This type of deterioration occurs more frequently in pollen grains and spores with thinner exine. In its extreme form, this can result in a condition where the sculpturing elements of exine become undefined, or apparently become a uniform mass, without structures. The degradation involves exposure of pollen grains and spores to the air, resulting in chemical oxidation. In peat bogs and lakes, pollen grains can undergo this type of decay due to dry periods.
- **Mechanical damages** cause ruptures breaks or creases in the exine, but they do not necessarily show reductions, thinning and perforations. The cause of this type of deterioration is usually the physical stress to which the pollen grains and spores were exposed in the course of their depositional process such as collisions due to fluvial transport; as the result of digestion observed in invertebrate coprolites whose pollen grains and spores are extremely wrinkled or even because of the compaction of sediments that may have occurred after their deposition.
- **Obscured pollen** grains and spores may be infiltrated with crystallized minerals *in situ*, or opaque debris may occur in the microscope slides affecting the visualization of palynomorphs.

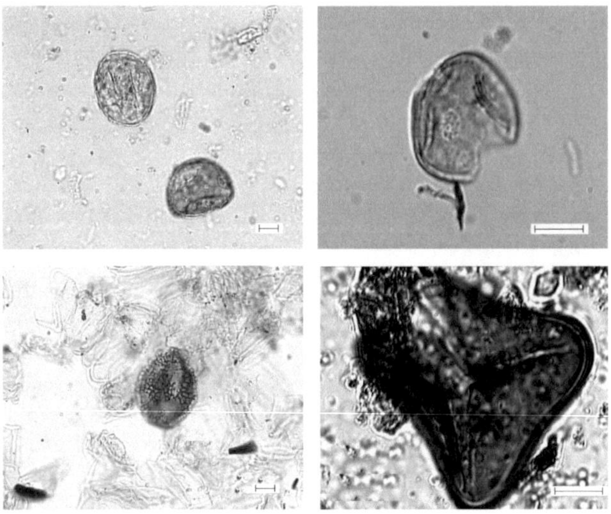

Fig. 7. Light photomicrographs of pollen and spore types of deterioration. Top: two examples of pollen corrosion. Bottom: examples of pollen degradation and of mechanical damage in a trilete fern spore. Scales in the figures= 10 µm.

Therefore, the pollen transport is different in different aquatic environments which will influence the overrepresentation or underrepresentation of certain pollen and spore *taxa* in the assemblage of the sediments. In general, there are:

- In **rivers** and **streams** a large amounts of pollen and spore can be transported by the currents and generally correspond to the vegetation of large surface areas where these currents flowed. Studies have shown that the sediments collected from within the main channel have a pollen assemblage representative of the distant mountains (regional elements) where the river passed by. Instead, the sediments adjacent to the main channel represent the vegetation near the sampling site (local elements).
- In **lakes**, during fluctuation, pollen and spores are subjected to the action of the winds along the water. Types of low sinking speed (smaller sizes) are preferentially deposited in the shallow areas of the coast. The sediment containing pollen, spores and algae can move to the deep parts of the lake ("sediment focusing"), below the region of wave action.
- In **deltas** and **estuaries** pollen and spores are transported by ocean currents and deposited together with the silt sediment. The pollen grains and spores carried by perennial rivers are deposited usually near the river mouth, in deltas. From the river mouth to the deeper parts of the ocean, the pollen grains and spores are deposited in decreasing gradients in relation of the grain size and concentration. In the spore-pollen assemblage, there can be high percentage of corroded, degraded and mechanically damaged pollen grains and spores. In estuaries, the mixing and reworking of sediments caused by the turbulence of the waves makes it almost impossible to pollen analysis. In general, you should be very careful when searching in marine environments because of the distortions in the palynomorphs spectra due to the complexity of the depositional patterns.

2. Sampling methods and analyses

Recent underwater deposition of pollen and spores can be studied by collecting surface sediment (first 2 cm) performed by equipment such as dredges, bottles, plastic tubes or with a modified free-fall valve corer (Davis *et al*. 1971). In collecting short cores, if there have not been rework or material loss, the top sediment corresponds to deposition of the last decades. It is important that collections of the surface sediments be carried out in a horizontal gradient ("transect") and along the direction of dominant winds, taking into account the direction of prevailing currents, including material at the margins as well as at the center of the depositional site. The number of samples is determined by the extent of the site. The chemical preparation employs a series of reagents in order to remove organic and inorganic residues in order to concentrate the palynomorphs in microscope slides (Faegri & Iversen 1950, Ybert *et al* 1992, among others). In order to assess the relative and absolute frequency of palynomorphs, pollen are counted, either by volume measurements (Cour 1974) or by introduction of exotic spores or pollen (Stockmarr 1971, Salgado-Labouriau & Rull 1986), making sure to perform observations in more than one slide, seeing both at the edges as at the center. It is common to count 300 grains of pollen per sample, but at the tropics, in general, this number needs to be larger in order to be noticed, even the rare grains (pollen types underrepresented in the sediment). The diagrams show the curves of Relative Frequency and Concentration of each palynomorph for each category and according to established Pollen-spore Sum, separating the regional elements from local elements which are indicators of humid environments. These diagrams may be plotted in different software programs (Polldata, Tilia, Coniss, C2, among others).

3. The influence of bathymetry associated with prevailing wind and fluvial currents - case study: Palynological depositional patterns in coastal lakes of the northern region of the Rio de Janeiro state, Brazil

The coastal plain of Campos dos Goytacazes municipality, northern coastal region of the Rio de Janeiro State, is an important area for palaeoenvironmental studies. This region presents several shallow lakes, which are relict bays of a large palaeolagoon system that was isolated from the sea during the Quaternary by sediments from the Paraíba do Sul River (e.g., Lagoa de Cima lake), by sand barriers or beach ridges (e.g., Lagoa Salgada lake), or by alluvial fans of the Barreiras Formation (e.g., Lagoa do Campelo lake).

The Lagoa de Cima Lake is embedded in a valley (Imbé River basin) located between the Barreiras Formation (Tertiary sediments) and the Precambrian crystalline basement, 50 km west from the coastal line and with about 30 m high. This lake may have been formed by an obstruction of a palaeolagoon called Ururaí Bay, and, therefore, represents the oldest lake in this region. The water is fresh and presents diatomite deposits at its margins. It is conditioned by the inflow of the Urubu and Imbé Rivers and presents an outlet called Ururaí that flows towards the Lagoa Feia Lake that is connected to the sea by a narrow passage. Nowadays, the Lagoa de Cima Lake drainage basin occupies an area of circa 986 km2 and does not present industrial activities but intense sugar-cane agriculture, pastureland, and a small remnant fragment of the Atlantic forest bordering the lake. The evergreen rainfall forest covers the high mountains of the drainage basin, especially inside the Parque Estadual do Desengano, a governmental area for the protection of the forest that is located 5 km west from the Lagoa de Cima Lake.

The Lagoa do Campelo Lake is located at 17 km away from the coastal line, with about 8 m high, bordering the Barreiras Formation and reaching the flattened sediments of the coastal plain, which cover the Cretaceous layers of the Campos Basin. Its drainage basin is not well limited and occupies an area of circa 98 km2. Without a tributary and an affluent, the lake receives fresh water and sediments from several swamps and bogs connected to the Paraíba do Sul River. The water of the lake was not naturally drained into the Atlantic Ocean because in 1950 the government carried out several changes in the Campos dos Goitacazes municipality in order to control the natural floods in this region. The building of a channel connecting this lake to the Paraíba do Sul River and another channel towards the sea was not good to its hydrological balance. A small remnant of the seasonal semideciduous forest can be observed at 5 km southwest of the lake, and a small swampy forest fragment of "restinga" vegetation in the northeastern margin of the lake. Pastureland, sugar-cane agriculture and subsistence plantations constitute the regional landscape of the drainage basin. The marsh vegetation at the lake borders presents Cyperaceae, Poaceae, some additional plant taxa, and a characteristic large belt of cattail (*Typha*).

Aiming to support the reconstruction of the temporal dynamics of the vegetation during the last 7,000 years (Luz *et al.* 2011), palynological studies of surface sediment samples were performed to elucidate the current dynamics that have influenced the sedimentation of palynomorphs inside these lakes. Fifteen surface samples were collected with a hand dredger in the top five centimeters of the Lagoa de Cima Lake sediments and four of the Lagoa do Campelo Lake, in a transect of 500 m steps from one to the opposite side, in the NE/SW direction, which is the same direction of the dominant wind (Luz *et al.* 2002, 2005, 2010).

Pollen grains occurring in samples obtained in the surface sediments from the transect across the Lagoa de Cima Lake reflect the extant vegetation around the lake and along the Imbé and Urubu Rivers, with an expressive contribution of arboreal regional taxa, as well as an important contribution of hydrophylous/swampy plant species and ruderal plants from the wide-ranging pasturelands in this region. In the Lagoa de Cima Lake, the ressuspended sediments (with corroded pollen and fern spores) take a preferential direction of deposition caused by water currents that generate high pollen rates in less deep and decentralized areas.

The smaller pollen grains and spores (5-25 µm) showed preferential deposition in the region of the water outlet of the lake (the Ururaí River) while the larger size grains, near the entrance of rivers Imbé and Urubu. The central region did not show a consistent pattern of deposition by grain size and presented also sand deposition at the most central point (sample 7) which prevented the preservation of palynomorphs in the sediment, making it sterile (Fig. 8).

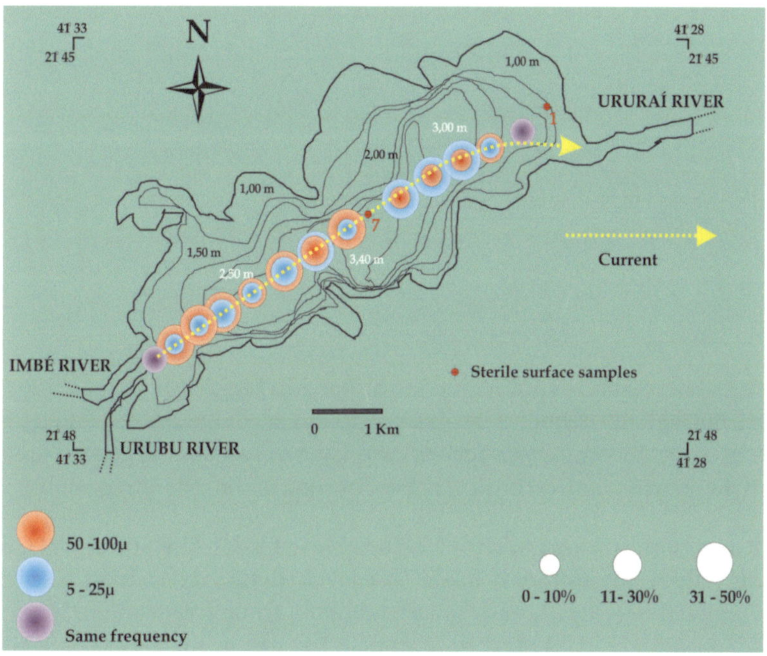

Fig. 8. Differential deposition of pollen and ferns spores in Lagoa de Cima lake, Rio de Janeiro, Brazil, through the frequency analysis of grain size.

Pollen and spores preservation was sometimes poor in the superficial sediments of the Lagoa do Campelo Lake, evidencing their exposition to the air during the partial drying of the lake and corrosion by microorganisms. The pollen spectra indicate a major performance of the local vegetation and the preferential deposition of regional pollen types at the southwest margin of the lake, reflecting the action of a dominant NE wind. The smaller pollen grains and spores (5-25 µm) showed preferential deposition in the central and deepest area of the transect, while the larger size grains were deposited preferentially near to the cattail belt at the north east margin of the lake (Fig. 9).

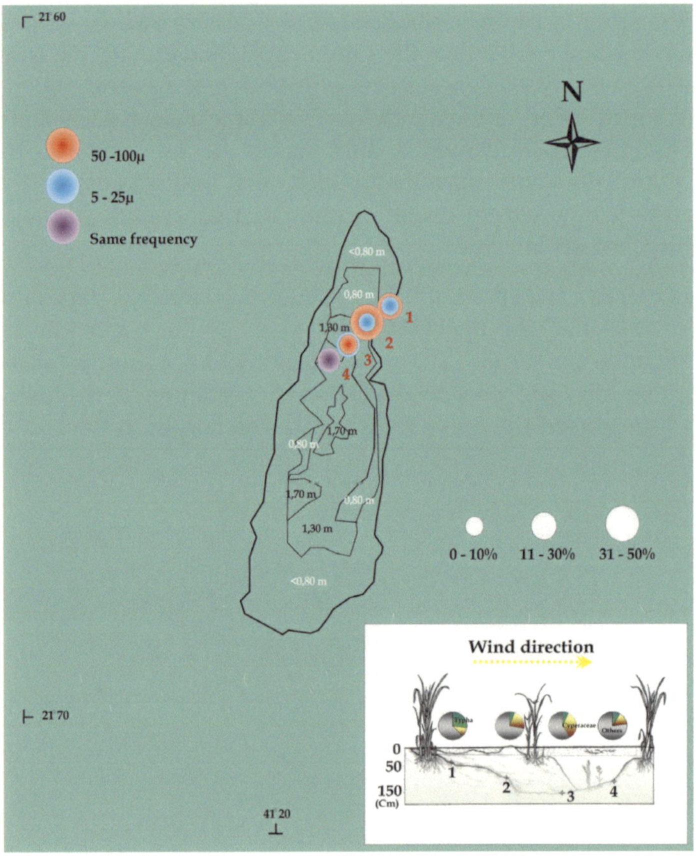

Fig. 9. Differential deposition of pollen and ferns spores in Lagoa do Campelo lake, Rio de Janeiro, Brazil, through the frequency analysis of grain size. In the right side image below: the pollen grains of autochthonous *Typha* (cattail) and Cyperaceae were deposited preferentially on the side of the source-plant, according to the direction of the prevailing wind.

The dynamic of deposition was different in the areas of the two studied lakes. At Lagoa de Cima Lake, it reflects the response to the sea level, always presenting a strong grouped influence of regional and local forest, grassland and swampy vegetation. Nevertheless, the dynamic of deposition in the Lagoa do Campelo Lake is in innermost dependence of the dominant wind and bathymetry. The high number of pollen types is attributed to the local plants.

4. The influence of bathymetry associated with tidal currents - case study: Palynological depositional patterns in Baia de Guanabara bay, Rio de Janeiro state, Brazil

The Guanabara Bay has a narrow entrance, approximately 1.6 km wide, which stretches towards S-N to its bottom, reaching a maximum diameter of 28 km, with a perimeter of 131 km. The water surface measures 373 km^2, excluding its islands and considering only its outer limits. Its basin covers approximately 4,600 km^2, including almost all the metropolitan areas of the municipalities of Rio de Janeiro and Niterói, among others. Around 35 rivers culminate into the bay and the longest ones (The Macacu and the Caceribu rivers) are born in the "Serra do Mar" mountain range. In the narrow entrance, there is a large sand bank, located at 22°56'48 "S/43°07'54" W, which rises from a depth of 20 m to 11 m. This feature promotes the canalling of the currents and acts as an obstacle to free movement of tidal currents. In the bottom topography of the bay, a feature that deserves mention is the center canal, general N-S orientation, stretching from the entrance neighborhood to near Ilha do Governador island. The most common depths of this canal are around 30 and 40 m, and near the Ilha Lage island, there is a depression which reaches 58 m of depth. The tides of the Guanabara Bay are classified as semi-daytime, with a period of about 12.5 h and differences in the high and low tides, whose amplitudes range from 0.20 to 1.40 m and with an average syzygy amplitude of about 1.20 m. The propagation of the tidal wave into its interior undergoes changes in phase and amplitude depending on the geometry of the Guanabara Bay.

The type of remaining vegetation in the region of the Guanabara bay is represented by the Rainforest (Atlantic forest domain), currently located in rugged topography (mountain slopes), mainly, and in a few forestry reserves. The areas of mangrove vegetation that, in the past, covered almost all of its edge, are currently limited to a continuous patch on the bottom of the bay and very sparse occurrences on its east coast.

A total of 27 surface samples were collected with a hand dredger in the top centimeters of the Guanabara Bay sediments (Barreto *et al.* 2006). The palynological results obtained showed a high percentage of herbaceous vegetation and a high richness of pollen grains from the mixed rainforest. The differential distribution of palynomorphs followed a pattern that was influenced by the bathymetry associated to the guidance of tidal currents, which originated the highest concentration of palynomorphs in the deeper and topographical obstructed areas (e.g., the islands) (Fig. 10).

Despite the complexity of currents that create resuspensions in the unconsolidated sediment surface in the entire Guanabara Bay and on the coast with an individualized morphology, differential deposition of palynomorphs followed a pattern of higher percentage of accumulation of small sizes at greater depths in the main canal, while the larger sizes preferably deposited in protected areas along the islands, laterally in lower depths and at the bottom of the bay after the discharge of major rivers. In the middle sector near Ilha do Governador island and at the bottom of the bay, concentrations of total palynomorphs were the highest in all samples analyzed. As the currents in the main canal near the mouth of the Baia de Guanabara Bay has a higher speed, it is assumed that in this area, the resuspended palynomorphs are constantly taken by the tides making their sedimentation difficult, which was observed in the analysis as a low total concentration of palynomorphs. This also explains the higher percentages of smaller sizes found there, once they are easier to be carried because they float longer. The islands with their points, coves and areas of less bathymetry favored the "capture" of palynomorphs in transit, resulting in a higher deposition there. The higher deposition of larger palynomorphs at the bottom of the bay, region of the culmination of several major rivers corroborates the idea of differential deposition in a gradient due to the grain size as postulated by several authors.

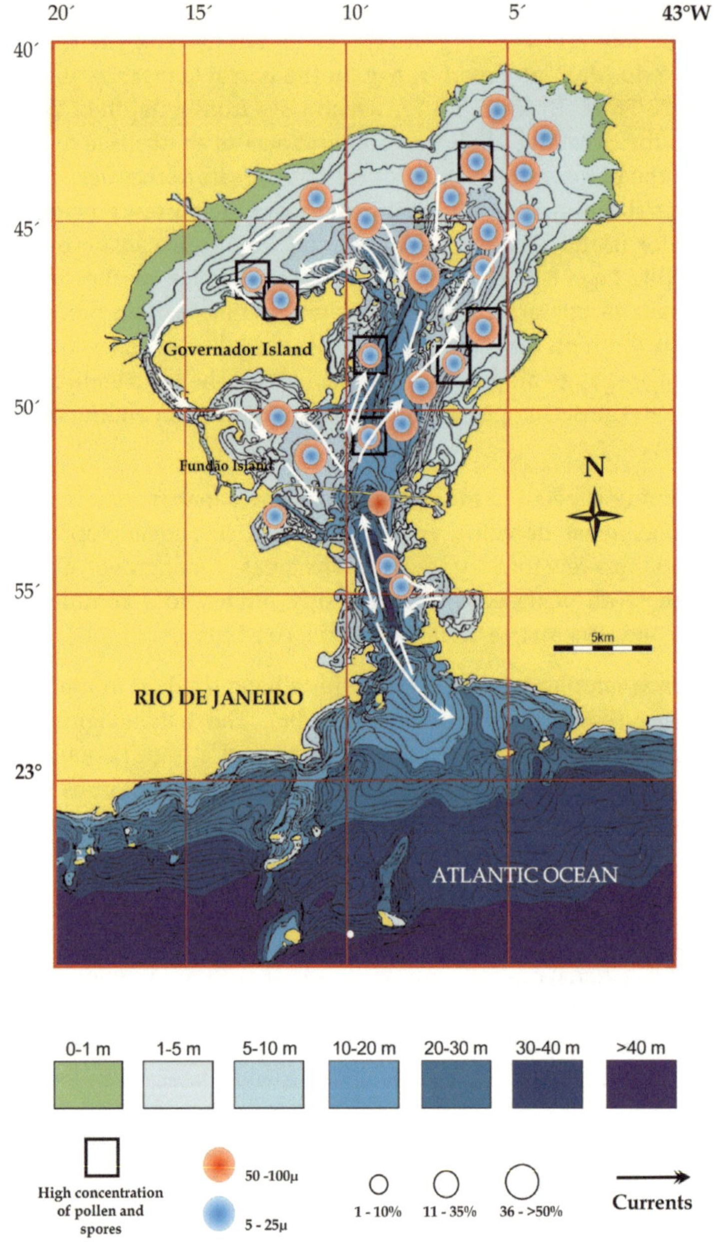

Fig. 10. Differential deposition of pollen and ferns spores in Baia de Guanabara bay, Rio de Janeiro, Brazil, through the frequency analysis of grain size.

5. Conclusions

The foundations of differential sedimentation in aquatic environments have been long investigated by palynologists, but their exact implications for the preservation of spore-pollen records are usually not fully taken into account in interpretations of changes in vegetation and climate through cores. The approach on the variable "space" in palynological research of the Quaternary is still neglected, both because there is a greater concern with the temporal dimension of vegetation obtained by the analysis of samples obtained at one single place and because there is usually no preliminary assessment of horizontal space of spore-pollen deposition in the study area. The lack of knowledge on this variable often leads to a poor choice of place for probing sample collection causing interpretation problems due to reversals of sedimentary layers and absence of palynomorphs deposition (Luz *et al.* 1996, Barth *et al.* 2001, Luz *et al.* 2011). Even between areas very close to each other or in two samplings obtained from the same site on a time-scale, disparate climate interpretations are seen through Palynology. The effects of this relation to local sedimentation problems of palynomorphs are often not explicitly addressed in the literature. However, in the last ten years, spatial analysis has been identified as a breakthrough in making ecology (hence, paleoecology) a more robust science (Pinto *et al.* 2003). Of course, factors such as differences in the pollen and spores productivity by the parent plant, modes of pollination and release of spores from each of them and their locations in the landscape with respect to the depositional site, present as significant problems to climate and paleoclimate reconstruction. However, one of the key points to understand is that in water transport, there is no uniform and continuous dispersion of palynomorphs, and they may suffer ressuspensions or temporary deposition beyond the sampling site, and may also go through various stages of wear and tear. This worn and broken material is an important environmental indicator and should be analyzed for its proportion in perfect and whole grains, because they provide important paleoecological information regarding temporal variations of the energy of water flows and possible droughts, exposing them to air. The disturbances in the sedimentation caused by hydrological and climatic changes alter both the quantity and the quality of the material deposited over time, such as interbedded sands amid packs of silt and clay, horizontal redistribution of previously deposited material causing reversals in the chronology of layers, higher occurrence of palynomorphs of large or small size at certain areas, etc., and should be taken into account especially where the compaction of the material is not yet big. An undisturbed laminar sedimentation is very rare in the tropics where there are no demarcated four seasons and where torrential rains are constant. Therefore, even if the Quaternary palynologist has a remarkable knowledge and is able to identify the family, genus or even the species to which the pollen and spore fossils belong, correct identification is only one side of the coin. Climate reconstruction over an image of the primitive vegetation that flourished in a given region through the palynology of sediments is a very difficult problem and, equally important is to know the possible causes of spatial variations in deposition of pollen grains, fern spores and algae, including undoubtedly the influence of bathymetry.

6. References

Andersen, S.T. (1970). The relative pollen productivity and pollen representation of North European trees, and correction factors for tree pollen spectra detrmined by surface pollen analyses from forests. *Danmarks Geologiske Undersøgelse,* Vol. *96, Serie II,* (january 1970), pp. 7–99

Barreto, C.F.; Barth, O.M.; Luz, C.F.P.; Coelho, L.G. & Vilela, C.G. (2006). Distribuição diferencial de palinomorfos na Baía de Guanabara, Rio de Janeiro, Brasil. *Revista Brasileira de Paleontologia*, Vol. 9, No. 1, (January 2006), pp. 117-126, ISSN 1519-7530

Barth, O.M.; Luz, C.F.P.; Toledo, M. B.; Barros, M.A. & Silva, C.G. (2001). Palynological data from Quaternary deposits of two lakes in the northern region of the state of Rio de Janeiro. In: *Proceedings of the IX International Palynological Congress. Houston*, Goodman, D.K. & Clarke, R.T., (Org.), pp. 443-450, American Association of Stratigraphic Palynologists Foundation, United States of America

Barth, O.M. 2004. Palinologia. In: *Paleontologia*, I.S. Carvalho, pp. 369-379, Editora Interciência, ISBN 85-7193-107-0, Rio de Janeiro, Brasil

Barth, O.M. & Melhem, T.S. (1988). *Glossário ilustrado de Palinologia*. Editora Unicamp, ISBN 85-268-0120-1, Campinas, Brasil

Bonny, A.P. (1980). Seasonal and annual variation over 5 years in contemporary airborne pollen trapped at a Cumbrian lake. *Journal of Ecology*, Vol. 68, No. 2, (July 1980), pp. 421–441, ISSN 00220477

Birks, H. H.; Whiteside, M. C.; Stark, D. M. & Bright, R. C. (1976). Recent paleolimnology of three lakes in northwestern Minnesota. *Quaternary Research*, Vol. 6, No. 2, (June 1976), pp. 249 - 272, ISSN 0033-5894

Bradshaw, R.H. (1981). Modern pollen representation factors for woods in south-east England. *Journal of Ecology*, Vol. 69, No. 1, (march 1981), pp. 45–70, ISSN 00220477

Bradshaw, R. H. & Webb, T. (1985). Relationships between contemporary pollen and vegetation data from Wisconsin and Michigan, USA. *Ecology*, Vol. 66, No. 3, (June 1985), pp. 721 - 737, ISSN 0012-9658

Bradley, W.H. (1965). Vertical density currents. *Science*, Vol. 150, No. 3702, (December 1965), pp. 1423-1428, ISSN 0036-8075

Brenner, W. & Foster, C.B. (1994). Chlorophycean algae from the Triassic of Australia. *Review of Palaeobotany and Palynology*, Vol. 80, No. 3-4, (February 1994), pp. 209–234, ISSN 0034-6667

Brooks, J. & Shaw, G. (1978). Sporopollenin: a review of its chemistry, palaeochemistry and geochemistry. *Grana*, Vol. 17, No. 2, (May 1978), pp. 91-97, ISSN 0017-3134

Calcote, R.R. & Davis, M.B. (1989). Comparison of pollen in surface samples of forest hollows with surrounding forests. *Bulletin of the Ecological Society of America*, Vol. *70*, No. supl., (January 1989), pp. 1-75, ISSN 0012-9623

Campbell, I.D. (1991). Experimental mechanical destruction of pollen grains. *Palynology*, Vol. 15, No. 1, (January 1991), pp. 29-33, ISSN 0191-6122

Campbell, I.D. (1999). Quaternary pollen taphonomy; examples of differential redeposition and differential preservation. *Palaeogeography, Palaeoclimatology, Palaeoecology*, Vol. 149, No. 1-4, (June 1999), pp. 245-256, ISSN 0031-0182

Catto, N.R. (1985). Hydrodynamic distribution of palynomorphs in a fluvial succession, Yukon. *Canadian Journal of Earth Science*, Vol. 22, No. 10, (October 1985), pp. 1552-1556, ISSN 0008-4077

Chaloner, W.G. (1976). The evolution of adaptative features in fossil exines. In: *The evolutionary significance of the exine*, Ferguson, I.K. & Muller, J., pp. 1-14, Academic Press, ISBN 0122536509, London

Chmura, G. L.; Smirnov, A. & Campbell, I. D. (1999) Pollen transport through distributaries and depositional patterns in coastal waters. *Palaeogeography, Palaeoclimatology, Palaeoecology,* Vol. 149, No. 1-4, (June 1999), pp. 257-270, ISSN 0031-0182

Colinvaux, P.A.; Frost, M.; Frost, I.; Liu, K.-B. & Steinitz-Kannan, M. (1988). Three pollen diagrams of forest disturbance in the western Amazon basin. *Review of Palaeobotany and Palynology,* Vol. 55, No. 1-3, (June 1988), pp. 73 - 81, ISSN 0034-6667

Cook, C.D.K. (1988). Wind pollination in aquatic angiosperms. *Annals of the Missouri Botanical Garden,* Vol. 75, No. 1, (January 1988), pp. 768–777, ISSN 0026-6493

Cour, P. (1974). Nouvelles techniques de detection des flux et des retombées polliniques: étude de Ia sédimentation des pollens et dês spores à Ia surface du sol. *Pollen et Spores,* Vol. 16, pp. 103-142, ISSN 0032-3616

Cross, A.T.; Thompson, G. G. E Zaitzeff, J. B. (1966) Source and distribution of palynomorphs in bottom sediments, southern part of Gulf of California. *Marine Geology,* Vol. 4, No. 6, (December 1966), pp. 467-524, ISSN 0025-3227

Cruz, N.M.C. (2004). Paleopalinologia, In: *Paleontologia,* I.S. Carvalho, pp. 381-3922, Editora Interciência, ISBN 85-7193-107-0, Rio de Janeiro

Cushing, E.J. (1967). Evidence for differential pollen preservation in late Quaternary sediments in Minnesota. *Review of Palaeobotany and Palynology,* Vol. 4, No. 1-4, (October 1967), pp. 87–101, ISSN 0034-6667

Davis, M.B. (1968). Pollen grains in lake sediments: redeposition caused by seasonal water circulation. *Science,* Vol. 162, No. 3855, (November 1968), pp. 796-799, ISSN 0036-8075

Davis, M.B.; Brubaker, L.B. & Beiswenger, J.M. (1971). Pollen grains in lake sediments: pollen percentages in surface sediments from southern Michigan. *Quaternary Research,* Vol. 1, No. 4, (December 1971), pp. 450–467, ISSN 0033-5894

Davis, M.B.; Moeller, R.E. & Ford, J. (1984). Sediment focusing and pollen analysis. In: *Lake sediments and environmental history,* Haworth, E.Y. & Lund, J.W.G., pp. 261-293, Leicester University Press, ISBN 0816613648, Leicester, England

Davis, M.B. & Brubaker, L.B. (1973). Differential sedimentation of pollen grains in lakes. *Limnology and Oceanography,* Vol. 18, No. 4, (July 1973), pp. 635-646, ISSN 0024-3590

Delcourt, P.A. & Delcourt, H. R. (1980) Pollen preservation and Quaternary environmental history in the southeastern United States. *Palynology,* Vol. 4, No.1, (January 1980), pp. 215-231, ISSN 0191-6122

Dias, I.C.A. (1997). *Chlorophyta filamentosas da Reserva Biológica de Poço das Antas, município de Silva Jardim, Rio de Janeiro: Taxonomia e aspectos ecológicos.* Thesis Universidade de São Paulo, Instituto de Biociências, 223 pp

Dupont, L.M. (1985). *Temperature and rainfall variation in a raised bog ecosystem: a palaeoecological and isotope-geological study.* Master Thesis, University of Amsterdam, 62 pp

Elsik, W.C. (1971). Microbial degradation of sporopollenin. In: *Sporopollenin,* Brooks J., Grant P.R., Muir M., Van Gijzel P. & Shaw G., pp. 480-511, Academic Press, New York, United States of America

Erdtman, G. (1943). *An introduction to pollen analysis,* Chronica Botanica Company, United States of America

Erdtman, G. (1952). *Pollen Morphology and Plant Taxonomy. An Introduction to Palynology* I. *Angiosperms* Almqvist & Wilksell, Stockholm

Erdtman, G. (1957). *Pollen and Spore Morphology and Plant Taxonomy. II. Gymnospermae, Pteridophyta, Bryophyta,* Almqvist & Wilksell, Stockholm

Erdtman, G. (1965). *Pollen and Spore Morphology and Plant Taxonomy. An Introduction to Palynology III. Gymnospermae, Bryophyta.* Almqvist & Wiksell, Stockholm

Erdtman, G. (1971). *Pollen and Spore Morphology and Plant Taxonomy. An Introduction to Palynology IV. Pteridophyta.* Almqvist & Wiksell, Stockholm

Erdtman, G. & Sorsa, P. (1971). *Pollen and spore. Morphology/plant taxonomy. Pteridophytas,* Almqvist & Wilksell, Stockholm

Faegri, K & Iversen, J. (1950). *Text book of modern pollen analysis,* Ejnar Munksgaard, Copenhagen

Faegri, K; Kaland, P.E. & Krzywinski, K. (1989). *Textbook of pollen analysis,* J. Wiley and Sons, ISBN 0-471-92178-5, New York, United States of America

Fall, P.L. (1987). Pollen taphonomy in a canyon stream. *Quaternary Research,* Vol. 28, No. 3, (November 1987), pp. 393-406, ISSN 0033-5894

Fritzsche, J. (1832). *Beiträge zur Kenntniss des Pollen,* Stettin & Elbing, Berlin

Graçano, D.; Prado, J. & Azevedo, A.A. (1998). Levantamento preliminar de pteridophyta do Parque estadual do Rio Doce (MG). *Acta Botanica Brasilica,* Vol. 12, No. 2, (April 1998), pp. 165–181, ISSN 0102-3306

Havinga, A.J. (1964). Investigation into the differential corrosion susceptibility of pollen and spores. *Pollen et Spores,* Vol. 6, pp. 621-635, ISSN 0375-9636

Heusser, L.E. (1978). Spores and pollen in the marine realm. In: *Introduction to marine micropaleontology,* Haq, B.U. & Boersma A., pp. 327–339, Elsevier Science, ISBN 0444002677, New York, United States of America

Heusser, L.E. & Balsam, W.L. (1977). Pollen distribution in the northeast Pacific Ocean. *Quaternary Research,* Vol. 7, No. 1, (January 1977), pp. 45-62., ISSN 0033-5894

Hopkins, J.S. (1950). Differential flotation and deposition of coniferous and deciduous tree pollen. *Ecology,* Vol. 31, No. 4, (October 1950), pp. 633–641, ISSN 00129658

Huszar, V. & Silva, L.H.S. (1992). Comunidades fitoplanctônicas de quatro lagoas costeiras do norte do Estado do Rio de Janeiro, Brasil. *Acta Limnologica Brasiliensia,* Vol. 4, No. 1, (January 1992), pp. 291-314, ISSN 2179-975X

Jacobson, G.L. & Bradshaw, R.H.W. (1981). The selection of sites for paleovegetational studies. *Quaternary Research,* Vol. 16, No. 1, (July 1981), pp. 80–96, ISSN 0033-5894

Jackson, S.T. (1990). Pollen source area and representation in small lakes of the northeastern United States. *Review of Palaeobotany and Palynology,* Vol. 63, No. 1, (May 1990), pp. 53–76, ISSN 0034-6667

Jackson, S.T. (1991). Pollen representation of vegetational patterns along na elevational gradient. *Journal of Vegetation Science,* Vol. 2, No. 5, (October 1991), pp. 613–624, ISSN 1100-9233

Jackson, S.T. (1994). Pollen and spores in quaternary lake sediments as sensors of vegetation composition: theoretical models and empirical evidence, In: *Sedimentation of organic particles,* Traverse, A., pp. 253–286, Cambridge University Press, ISBN 9780521675505, Cambridge, England

Jackson, S. T.; Futyma, R. P. & Wilcox, D. A. (1988). A paleoecological test of a classical hydrosere in the lake Michigan Dunes. *Ecology,* Vol. 69, No. 4, (August 1988), pp. 928-936, ISSN 0012-9658

Jansonius, J. & Mcgregor, D.C. (1996). *Palynology: Principles and applications*, American Association of Stratigraphic palynologists Foundation, ISBN 0-931871-03-4, United States of America

Janssen, C.R. (1966). Recent pollen spectra from the deciduous and coniferous-deciduous forests of northeastern Minnesota: a study in pollen dispersal. *Ecology*, Vol. 47, No.5, (September 1966), pp. 804–825, ISSN 0012-9658

Janssen, C.R. (1967). A comparison between the recent regional pollen rain and the sub-recent vegetation in four major vegetation types in Minnesota (U.S.A.). *Review of Palaeobotany and Palynology*, Vol. 2, No. 1-4, (June 1967), pp. 331–342, ISSN 0034-6667

Janssen, C.R. (1973). Local and regional pollen deposition. In: *Quaternary Plant ecology*, Birks, H.J.B. & West, R.G., pp. 31–42, Blackwell, ISBN 0632091207, Oxford, England

Kawase, M. & Takahashi, M. (1995). Chemical composition of sporopolenin in Magnolia grandiflora (Magnoliaceae) and Hibiscus syriacus (Malvaceae), *Grana*, Vol. 34, No. 4, (October 1995), pp. 242-245, ISSN 0017-3134

Kershaw, A.P. & Hyland, B.P.M. (1975). Pollen transfer and periodicity in a rain-forest situation. *Review of Palaeobotany and Palynology*, Vol. 19, No. 2, (March 1975), pp. 129–138, ISSN 0034-6667

Leelinger, D.B. (2002). *A modern multilingual glossary for taxonomic pteridology*. American Fern Society, ISBN 0-933500-02-5, United States of America

Lehman, J.T. (1975). Reconstructing the rate of accumulation of lake sediment: the effect of sediment focusing. *Quaternary Research*, Vol. 5, No. 4, (December 1975), pp. 541–550, ISSN 0033-5894

Lorscheitter, M.L. (1989). Palinologia de sedimentos quaternários do testemunho T15, cone Rio Grande, Atlântico Sul, Brasil. Descrições taxonômicas. Parte II. *Pesquisas*, Vol. 22, No. 1, (January 1989), pp. 89–127, ISSN 0373-840X

Luz, C.F.P.; Barth, O.M.; Martin, L.; Turcq, B. & Flexor, J.M. (1996). Estudos Palinológicos na Lagoa de Cima, Rio de Janeiro, Brasil. In: *Symposium Dynamique à Long Terme des Écosystèmes Forestiers Intertropicaux*, pp. 239-242, Unesco-Mab-Ird Bondy, France

Luz, C.F.P.; Nogueira, I.S.; Barth, O.M. & Silva, C.G. (2002). Differential Sedimentation of Algae Chlorococcales (*Scenedesmus*, *Coelastrum* and *Pediastrum*) in Lagoa de Cima, Campos dos Goitacazes Municipality (Rio de Janeiro, Brazil). *Pesquisas em Geociências*, Vol.29, No.2, (April 2002), pp. 65-75, ISSN 1518-2398

Luz, C.F.P.; Barth, O.M. & Silva, C.G. (2005). Spatial distribution of palynomorphs in the surface sediments of the Lagoa do Campelo lake, North region of Rio de Janeiro State, Brazil. *Acta Botanica Brasilica*, Vol. 19, No. 4, (October 2005), pp. 741-752, ISSN 0102-3306

Luz, C.F.P.; Barth, O.M. & Silva, C.G. (2010). Modern processes of palynomorph deposition at lakes of the northern region of the Rio de Janeiro State, Brazil. *Anais da Academia Brasileira de Ciências*, Vol. 82, No. 3, (September 2010), pp. 679-690, ISSN 0001-3765

Luz, C.F.P.; Barth, O.M.; Martin, L.; Silva, C.G. & Turcq, B.J. (2011). Palynological evidence of the replacement of the hygrophilous forest by field vegetation during the last 7,000 years B.P. in the northern coast of Rio de Janeiro, Brazil. *Anais da Academia Brasileira de Ciências*, Vol. 83, No. 3, (September 2011), pp. 939-951, ISSN 0001-3765

Margalef, R. (1969). Comunidades planctônicas em lagunas litorales. In: *Lagunas costeras, un simposio*, Castañares, A.A. & Phleger, F.B., pp. 545–562, Unam - Unesco, México

Melhem, T.S.; Giulietti, A.M.; Forero, E.; Barroso, G. M.; Silvestre, M.S.F.; Jung, S.L.; Makino, H.; Melo, M.M.R.F.; Chiea, S.C.; Wanderley, M.G.L.; Kirizawa, M.& Muniz, C. (1981). Planejamento para elaboração da "Flora Fanerogâmica da Reserva do Parque Estadual das Fontes do Ipiranga (São Paulo, Brasil)". *Hoehnea,* Vol. 9, No. 1, (January 1981), pp. 63-74, ISSN 0073-2877

Merilainen, J. (1969). Distribution of diatom frustules in recent sediments of some meromictic lakes. *Mitteilungen - Internationale Vereinigung fur Theoretische und Angewandte Limnologie,* Vol. 17, pp. 186–192, ISSN 0538-4680

Moore, P.D.; Webb, J.A. & Colinson, M.E. (1991). *Pollen Analysis.* Second Edition. Blackwell Scientific Publications, ISBN 0-632-02176-4, Oxford, England

Muller, J. (1959). Palynology of recent Orinoco delta and shelf sediments. *Micropaleontology,* Vol. 5, No. 1, (January 1959), pp. 1-32

Parsons, R. W., Prentice, I. C. & Saarnisto, M. (1980). statistical studies on pollen representation in Finnish lake sediments in relation to forest inventory data. *Annales Botanici Fennici,* Vol. 17, No. 4, (October 1980), pp. 379–393, ISSN 0003-3847

Parsons, R.W. & Prentice, I.C. (1981). Statistical approaches to R-values and the pollen-vegetation relationship. *Review of Palaeobotany and Palynology,* Vol. 32, No. 2-3, (March 1981), pp. 127–152, ISSN 0034-6667

Peck, R.M. (1973). Pollen budget studies in a small Yorkshire catchment. In: *Quaternary Plant Ecology,* Birks, H.J.B. & West, R.G., pp. 43–60, Blacwell Science Ltd, ISBN 0632091207, Oxford, England

Pennington, W. (1996). Limnic sediments and the taphonomy of late glacial pollen assemblages. *Quaternary Science Review,* Vol. 15, No. 5-6, (December 1996), pp. 501-520, ISSN 0277-3791

Pinto, M. P.; Bini, L. M. & Diniz-Filho, J. A. F. (2003). Análise quantitativa da influência de um novo paradigma ecológico: autocorrelação espacial. *Acta Scientiarum: Biological Sciences,* Vol. 25, No. 1, (January 2003), pp. 137-143, ISSN 1679-9283

Potter, L.D. (1967). Differential pollen accumulation in water-tank sediments and adjacent soils. *Ecology,* Vol. 48, No. 6, (November 1967), pp. 1041-1043, ISSN 0012-9658

Prentice, I.C. (1985). Pollen representation, source area, and basin size: toward a unified theory of pollen analysis. *Quaternary Research,* Vol. 23, No. 1, (January 1985), pp. 76–86, ISSN 0033-5894

Prentice, I.C. (1988). Records of vegetation in time and space: the principles of pollen analysis. In: *Vegetation History,* Huntley, B. & Webb, T., pp. 17–42, Kluwer Academic Publishers, ISBN 9061931886, Dordrecht, Netherlands

Prentice, I. C.; Berglund, B. E. & Olsson, T. (1987). Quantitative forest-composition sensing charcteristcs of pollen samples from Swedish lakes. *Boreas,* Vol. 16, No. 1, (March 1987), pp. 43–54, ISSN 0300-9483

Punt, W.; Hoen. P.P.; Blackmore, S.; Nilsson, S. & Le Thomas, A. (2007). Glossary of pollen and spore terminology (second edition), *Review of Palaeobotany and Palynology,* Vol. 143, No. 1-2, (January 2007), pp. 1-81, ISSN 0034-6667

Roubik, D.W. & Moreno, P.J.E. (1991). *Pollen and spores of Barro Colorado Island,* Missouri Botanical Garden, ISSN 0161-1542, United States of America

Rowley, J.R. (2001). Why the endexine and ectexine differ in resistance to oxidation. *Calluna* as a model system. *Grana,* Vol. 40, No. 3, (October 2001), pp. 159–162, ISSN 0017-3134

Salgado-Labouriau, M.L. (1973). *Contribuição a Palinologia dos Cerrados.* Editora Academia Brasileira de Ciências, Rio de Janeiro, Brazil

Salgado-Labouriau, M.L. (2007). *Critérios e técnicas para o Quaternário.* (first edition), Editora Edgard Blucher, ISBN 85-212-0387-X, São Paulo, Brazil

Salgado-Labouriau, M.L. & Schubert, C. (1977). Pollen analysis of a peat bog from Laguna Victoria (Venezuelan Andes). *Acta Científica Venezolana,* Vol. 28, No.3, (July 1977), pp. 328-332, ISSN 0001-5504

Salgado-Labouriau, M.L. & Rull, V. (1986). A method of introducing exotic pollen for palaeoecological analysis of sediments. *Review of Paleobotany and Palynology,* Vol. 47, No. 1-2, (February 1986), pp. 97-103, ISSN 0034-6667

Stockmarr, J. (1971). Tablets with spores used in absolute pollen analysis. *Pollen et Spores,* Vol. 13, No. 14, pp. 615-621, ISSN 0032-3616

Stommel, H. (1949). Trajectories of small bodies sinking slowly through convection cells. *Journal of Marine Research,* Vol. 8, No. 1, (January 1949), pp. 24–29, ISSN 0022-2402

Sugita, S. (1993). A model of pollen source area for na entire lake surface. *Quaternary Research,* Vol. 39, No. 2, (March 1993), pp. 239–244, ISSN 0033-5894

Tauber, H. (1965). Differential pollen dispersion and the interpretation of pollen diagrams. *Danmarks Geologiske Undersøgelse,* Vol. 89, No. 2, (April 1962), pp. 1–69, ISSN 0366-9130

Tauber, H. (1977). Investigations of aerial pollen transport in a forested area. *Dansk Botanisk Arkiv,* Vol. 32, No. 1, (January 1977), pp. 1–121, ISSN 0011-6211

Traverse, A. (1988). *Paleopalynology* (second edition), Allen & Unwin Hyman, ISBN 0045610010, Boston, United States of America

Traverse, A. & Ginsburg, R.N. (1966). Palynology of the surface sediments of Great Bahama Bank, as related to water movement and sedimentation. *Marine Geology,* Vol. 4, No. 6, (December 1966), pp. 417-459, ISSN 0025-3227

Tryon, A.F. & Lugardon, B. (1990). *Spores of the Pteridophyta. Surface, wall structure, and diversity based on electron microscope studies*, Springer- Verlag New York Inc, ISBN 0-387-97218-8, New York, United States of America

Tschudy, R.H. & Scott, R.A. (1969). *Aspects of Palynology,* Wiley-Interscience, ISBN 0471892203 9780471892205, New York, United States of America

van Geel, B. (1976). *A paleoecological study of Holocene peat sections based on the analysis of pollen, spores and macro and microscopic remains of fungi, algae, cormophytes and animals.* Thesis University of Amsterdam, Hugo de Vries Laboratorium, 75 pp

van Geel, B. & van Der Hammen, T. (1978). Zygnemataceae in Quaternary Colombian sediments. *Review of Palaeobotany and Palynology,* Vol. 25, No. 5, (June 1978), pp. 377-392, ISSN 0034-6667

von Post, L. (1916). Om skogstradspollen I sydsvenska torfmosselagerfoljder (foredragsreferat). *Geologiska Föreningens Stockholm Förhandlingar,* Vol. 38, pp. 384-394, ISBN 1149146842

von Post, L. (1967). Forest tree pollen in South Swedish peat bog deposits. *Pollen et Spores,* Vol. 9, No. 3, pp. 375-401, ISSN 0032-3616

Wang, K.; Zhang, Y. & Sun, Y. (1982). The spore-pollen and algae assemblages from the surface layer sediments of the Yangtze River delta. *Acta Geographica Sinica,* Vol. 37, No. 3, (July 1982), pp. 261-271, ISSN 0375-5444

Webb, T.; Yeracaris, G.Y. & Richard, P. (1978). Mapped patterns in sediments samples of modern pollen from southeastern canada and northeastern united States. *Geographie Physique et Quaternaire,* Vol. 32, No. 2, (April 1978), pp. 163–176, ISSN 0705-7199

Wodehouse, R.P. (1935). *Pollen grains – their structure, identificationand significance in science and medicine,* McGraw-Hill Publ., New York, United States of America

Whitehead, D.R. (1969). Wind pollination in the angiosperms evolutionary and environmental considerations. *Evolution,* Vol. 23, No. 1, (January 1969), pp. 28–35, ISSN 1558-5646

Whitehead, D.R. (1983). Wind pollination: some ecological and evolutionary perspectives. In: *Pollination Biology,* Real, L., pp. 97–108, New York Academic Press Inc, ISBN 0125839820, New York, United States of America

Ybert, J.P.; Salgado-Labouriau, M.L.; Barth, O.M.; Lorscheitter, M.L.; Barros, M.A.; Chaves, S.A.M.; Luz, C.F.P.; Ribeiro, M.; Scheel, R. & Vicentini, K.F. (1992). Sugestões para padronização da metodologia empregada em estudos palinológicos do Quaternário. *Revista Instituto Geológico São Paulo,* Vol. 13, No. 2, (July 1992), pp. 47-49

Zetzsche, F. (1932). Die sporopollenine, In: *Handbuch der pflanzenanalyse,* G. Klein, pp. 205-215, Springer-Verlag, Viena